FISHING
AROUND
MORECAMBE BAY

FISHING
AROUND
MORECAMBE BAY

MIKE SMYLIE

For Moe, Ana & Otis… with love.

First published 2010

The History Press
The Mill, Brimscombe Port
Stroud, Gloucestershire, GL5 2QG
www.thehistorypress.co.uk

© Mike Smylie, 2010

British Library Cataloguing in Publication Data.
A catalogue record for this book is available from the British Library.

ISBN 978 0 7524 5393 4

Typesetting and origination by The History Press
Printed in Great Britain
Manufacturing managed by Jellyfish Print Solutions Ltd

CONTENTS

ACKNOWLEDGEMENTS

Several people have contributed towards the final presentation of this book. Primarily, Jennifer Snell of Ulverston has helped the most, with both the *Hearts of Oak* and Flookburgh shrimping. Many photographs from her collection have been used in these two chapters. Thanks must go to Brian Hampson and Mike Arridge for their photos of the *Hearts of Oak*. In Fleetwood, Phil at Phil's Fylde coast post-card website – www.rossallbeach.co.uk – allowed me access to many of his photos, for which I am grateful, and for his quick response to my initial email. Many thanks to Tom Smith for suffering several visits, interrupting his busy work, during which he recounted various aspects of his life on the low shore. The homemade cakes were the best ever! Lancaster Maritime Museum has supplied many photographs, some of which came from Keith Willacy who allowed permission for their use. Thanks to Len Lloyd for the loan of four photos, even if it was a few years ago, and to Peter Brady for introducing me to Frank Clarkson and Dick Massey. For places to visit, Morecambe Bay has some excellent maritime museums. A visit to all is to be recommended, though opening hours should be checked: Fleetwood Museum (01253 876621), The Dock Museum, Barrow (01229 876400), Lancaster Maritime Museum (01524 382264), Fleetwood Maritime Heritage Trust (01253 872219) and Heysham Heritage Centre (no telephone, run by volunteers).

Non-credited photographs were taken by the author, or are from his collection.

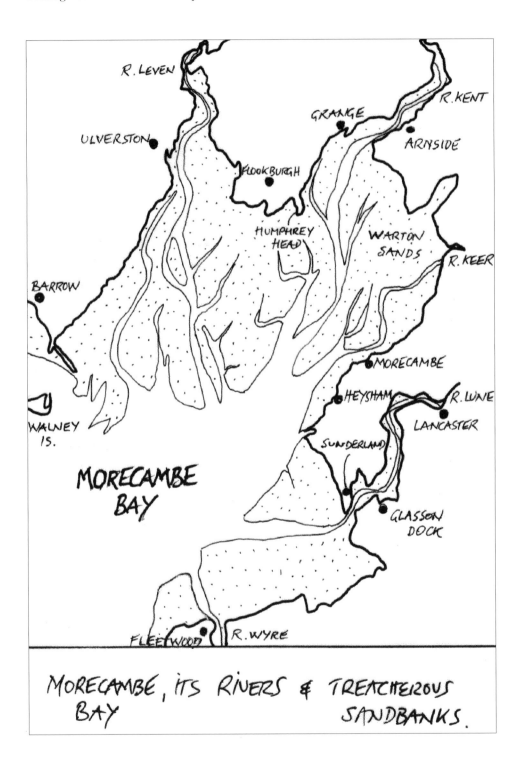

MORECAMBE, ITS RIVERS & TREACHEROUS BAY SANDBANKS.

INTRODUCTION

When Daniel Defoe came to Lancaster in the early eighteenth century he described that part of the country as 'very strange'. As to the surrounding sea, this was 'itself desolate and wild, for it was a sea without ships, here being no sea-port or place of trade... the people told us they should not see a ship under sail for many weeks together'. Lancaster itself he regarded as having little to recommend it 'but for a decayed castle, and a more decayed port'. No ships of any considerable burthen used it as far as he was concerned. Morecambe Bay did not merit much attention. Apparently he missed so much, though it must be said the coasts of Lancashire (including Furness which comprised much of the north and west of the bay and which was considered to be an exclave of the historic county of Lancashire at the time) and Westmorland were sparsely inhabited. He did, however, mention the existence of Cartmel Sands which, it is thought, might have been the old name for Morecambe Bay. Even today Cartmel Sands and Cartmel Wharf together cover the vast majority, though not the entirety, of the bay.

Morecambe Bay is Britain's second biggest bay, only beaten by the Wash. Like the Wash, it is rich in many species of fish, especially shellfish. It is a bay open to the south-west, some 12 miles wide by 18 miles long, covering an area a little over 121 square miles. It is the biggest continuous intertidal area in Britain, one reason it has such vast supplies of shrimps, for which it has been famous for many years. Around its circumference it has 5 per cent of the UK's total area of salt marsh. The tidal range is huge at over 33ft during spring tides, and the ebb tide can, in places, retreat nearly 8 miles. At that point the bay is a vast eerie area of sand sometimes known as the 'Wet Sahara'. When the tide floods it can come in 'quicker than the speed of a trotting horse' and tidal bores can reach nine knots and cover huge tracts of sand flats in minutes. Given that areas are also subject to quicksands, it is not surprising that lives often get lost out on the sands.

Several rivers flow into the bay – the rivers Kent, Keer, Leven, Lune and Wyre – with the river Kent being one of the fastest flowing rivers in England. Thus the salinity of the bay is lower than the neighbouring seas. Towns such as Fleetwood, Heysham, Morecambe, Grange-over-Sands, Ulverston and Barrow-in-Furness have grown up around its edge. In the nineteenth century it was through a flourishing tourism industry that working people came to this coast, leaving an imprint that still affects the area today. Several industries, from shipbuilding to nuclear power generation, keep some 200,000 people who live and work around its shores employed. Fishing, probably amongst the oldest activity around the bay, still plays a very small part in the economy of the area.

Lancaster, a century after Defoe passed, was a hive of activity. By then it had a canal that connected it to the hinterland and much further away – 500 miles further away! Sail cloth was one of the city's chief manufacturing products and there were ropewalks, both commodities badly needed by fishermen and sailors alike. Ships were also built in Lancaster. In 1801 there were forty-seven vessels involved in trade with the West Indies. However, with regard to the West Indies, many of these ships – some built in Lancaster – were involved in the slave trade, and Lancaster has the dubious honour of being the fourth largest slaving port in the country. Over 200 journeys were made by Lancaster vessels, carrying over 250,000 Africans to the West Indies and Americas to be sold as slaves, contributing to this human misery. Money from this trade flowed into the city; buildings were built and Lancaster flourished from its immoral earnings. However, there was, in all reality, no benefit to the working people of the surrounding area. Fishing continued on a subsistence level until the end of the nineteenth century and into the twentieth.

Now at the dawn of the twenty-first, fishing is almost back to pre-Defoe levels. Nevertheless it has left behind a strong tradition of man's determination to gather harvest of every possible nature from the wild and treacherous sea. Many different methods have been used over the centuries which today leave us with a rich history to discover. Baulk hedges and stake nets once adorned the sands, cockles and mussels were there for the picking, whilst various netting techniques continue to be used today such as whammel-nets, drift-nets, seine-nets, haff-nets, push-nets and even beam trawls dragged behind tractors for shrimps. Others, meanwhile, used hooks and spears to scratch a meagre living from the sands.

Added to this diversity is the town of Fleetwood, from where boats sailed out to Iceland and other northerly latitudes returning laden with fish, and which became one of the biggest ports of trawled fish in Britain. Morecambe Bay, therefore, is a perfect representation of some of the varied fishing methods practised throughout Britain, and some of its history is thus presented within these pages.

ONE

THE DEVELOPMENT OF FLEETWOOD AND ITS DEEPSEA TRAWLERS

Fleetwood today seems to have a grey cloud suspended over it, a place that does not exactly fit into the twenty-first century. Somehow it seems to be a town of ghosts, of lost aspirations and cheated goals, due partly from the loss of Icelandic fishing grounds and partly from the European Common Fisheries Policy. Today the once-great fish dock only has a handful of working vessels – four trawlers and a couple of potters – as well as the 'heritage trawler' *Jacinta* and several de-registered boats, and even the adjacent marina is only half full.

Jubilee Quay, once that bastion of free landing, is not free any longer. It is owned by Associated British Ports and signs around the commercial harbour talk of security risks and threaten prosecution under regulation (EC) No.725/2004, which is not that surprising seeing as ABP own that as well and they do not have a good public relations profile in many quarters.

What were once the fish processing buildings are now home to The Fleetwood Freeport, an outlet shopping area where nothing is free, which makes one wonder why it is called a Freeport. At the same time most of the shops, some full of items at knock-down prices, look as if they are part of a film set. Maybe they are in some way representing today's culture of imported goods from the East rather than yesterday's fish from the northern seas. What is ironic is that, on exiting the shopping area, there is a distinct smell of fish! With its southern neighbour, Blackpool, outdoing it in every field, and the glory of its past hanging over it, Fleetwood is a shadow of its former self – a great shame as it does not deserve that.

Fleetwood was, throughout much of the twentieth century, a humdinger of a fisheries port, responsible for sending fleets of trawlers northwards to Iceland and beyond. However, before the end of the last decade of the nineteenth century it was simply another base for inshore fishing boats working the nearby Irish Sea,

Margaret, FD208, and a nobby passing each other on the river Wyre. The *Margaret* was the first of Fleetwood's boats to have a petrol motor installed. (Photo courtesy of Sankeys of Barrow)

This time two trawlers – *Wonder*, FD68, and *Harriet*, FD111 – are sailing downriver and out to sea, their beam trawls clearly visible over the counter stern. (Photo courtesy of Sankeys of Barrow)

Opposite from top
A bustling quayside at Fleetwood as the lumpers and labourers move boxes and barrels of fish along. (Photo courtesy of www.rossallbeach.co.uk)

A fine view of the trawler *Desdemona*, FD203, as she sails past a steam ship. (Photo courtesy of www.rossallbeach.co.uk)

The Wyre Light in the 1920s, a welcome sight after three weeks or more at sea. (Photo courtesy of www.rossallbeach.co.uk)

Fleetwood-registered fishing smacks alongside Jubilee Quay. (Photo courtesy of www. rossallbeach.co.uk)

The trawler *Westlyn*, FD8, built in Middlesbrough in 1914 as the *City of Selby* and typical of her time. She was eventually scrapped in 1959. (Photo courtesy of www.rossallbeach.co.uk)

whilst over half a century before that, in the 1830s, it was, according to Samuel Lewis, nothing but a rabbit warren and a solitary lime kiln.

Thus nothing in Fleetwood is old in contemporary terms. Even the oldest pub in town, the Victoria, does not look very old. With the town's surprise rise in popularity among the ranks of the fleet owners came its sudden jump towards the head of the table in terms of fishery landings, third in Britain as a whole.

Like much of that part of the Lancashire coast, the area around the estuary of the river Wyre was virtually uninhabited at the beginning of the nineteenth century. Between Blackpool, where tourism was starting to develop, and Glasson Dock, there was little more than sea, sand and grass inhabited by rabbits and sea birds. The land itself was owned by the Hesketh family of Rossall Hall, near Rossall Point, which is considered to be the southern corner of Morecambe Bay, the point where the bay meets the Irish Sea.

In fact, the Hesketh family originated from North Meols until, in 1733, Roger Hesketh married Margaret Fleetwood, whose family owned Rossall Hall. Following the union the two families controlled most of the land between Rossall and North Meols. By the beginning of the nineteenth century, Bold Hesketh, son of Roger and Margaret, and by then Lord of the Manor, had a small fleet of fishing boats, supplying not only his estate but the surrounding villages.

When not in use, the boats were beached on the sandy expanse of Rossall Beach, though various storms were known to have wrecked several of them over the years. Within a decade these had been moved to moorings in the river Wyre where a few fishermen's huts were built at the same time, though these were obviously missed by Lewis when he visited, assuming of course that he did. Either that or they were demolished after Bold died in 1819 and his younger brother Robert became the new lord. Five years later Robert's fifth son Peter had inherited the estate, his four elder brothers having died prematurely.

Peter has been described as a bit of a radical who supported co-operative industrial schemes and other ideas not generally regarded as 'proper' in the eyes of the majority of the landed gentry. After the inaugural railway opening between Liverpool and Manchester in 1830 to which he had been invited, Peter had a vision of building a new town for visitors to be accommodated in the summer, a town modelled on St Leonards-on-Sea where he had spent many a holiday. The visitors would be drawn from the booming Lancashire mill towns, shipped in by railway, whilst to create winter work for the permanent inhabitants he was persuaded to incorporate a port that would also ease the burden from overloaded and over-priced Liverpool.

The following year Peter changed his surname, by Royal assent, to Hesketh-Fleetwood, in respect of his maternal grandmother's ancestors. Later that year the Preston & Wyre Railway Company was formed to build a new track to bring in the necessary building materials, and, later, the holiday-makers and day-trippers. The new town was, rather unsurprisingly, to be called Fleetwood, and in 1836 the first building was erected. Thus by the end of the decade the first inhabitants had moved in. Presumably Lewis failed to note the proposed development!

By 1840 a pilot service had been stationed in the Lune Deeps to help vessels proceeding up the Wyre channel to the small ports at Skippool and Wardleys, where cargoes such as flax, tallow and grain had been imported for centuries. The port of Fleetwood was also well underway and the Customs House had already been moved there in 1838. Whilst waiting for incoming vessels, the pilots often fished from their small cutter using a confiscated trawl net that customs officer Robert Roskell, himself the son of a fisherman, had taken from a smuggling boat. Before long they were making more money from the fishing than the pilotage and a consortium of three men (some say Roskell himself was a member) started using the *Pursuit*, by which time it had been retired as the pilot boat, and established the Fleetwood Fishing Company.

In 1841 the consortium hired four small smacks for the fishing season from the Leadbetter family of Banks, near Southport, a fishing family with a tradition of inshore fishing. These proved highly successful and five further half-decked smacks were purchased the following year, these being an early form of the type that was to become known as the Lancashire or Morecambe Bay nobbies. However, the enterprise only lasted a few years before they packed up, selling the boats.

Robert Roskell, however, having observed their success, began himself to use the *Pursuit*, which had previously been laid up, and the same trawl net. In 1846 Roskell was joined by Kirkcudbright fisherman John Wright, who moved his family and his smack to Fleetwood. Four years later the Leadbetters, finding fishing unprofitable off

Fishermen aboard, resting after hauling the net. Fishing was one of the hardest, and certainly one of the most dangerous, occupations. (Photo courtesy of www. rossallbeach.co.uk)

'Couch' Wright in his 1893-built Crossfield's nobby, *Charlotte*, on his way past Jubilee Quay.

The *Wyre Conqueror*, FD187, a modern trawler built in 1956 at Beverley and one of many belonging to Wyre Trawlers Ltd. She was scrapped in 1979. (Photo courtesy of www. rossallbeach.co.uk)

The *Mercurius*, FD277, built in Holland in 1963, alongside at Milford Haven in 2009.

the Ribble Estuary, moved to Fleetwood and the town became established as a fishing station. Finally another two families, coincidentally also named Wright, moved in, one from Marshside, Southport and the other from the east coast. The main catch seems to have been plaice, skate, codling and haddock, with herring being landed in the season.

By 1851 the population of Fleetwood had reached over 3,000. Regular fish sales were being held and fish storage sheds had been erected to cope with the increasing amount of fish that was being landed as the fleet of small fishing boats grew. Most of the fish was taken within the bay or just outside, and with the railway having been completed in 1840 much was taken away to the Lancashire towns. At the same time the railway was bringing in thousands of visitors. An advertisement of the 1850s shows fares for Sunday mornings when early trains allowed visitors to 'bathe and refresh themselves in ample time to attend a Place of Worship'.

By 1860 there were thirty-two small vessels working from the port and this number had doubled by 1876, when 100 tons of fish was being shipped out each day. Many of the boats had been built by boatbuilders working from the beach or along Dock Street, whilst others came from Freckleton on the Ribble, nearby Glasson Dock and Arnside on the river Kent. A typical smack cost between £300 and £500, with a down payment of £100 and the rest being paid as the boat made money.

It is presumed that the majority of these vessels were also of the nobby type. However, as the fleet grew, boats were brought in from Brixham, Rye and Great

Stern view of the stern trawler *Jacinta* at her permanent berth in the Wyre Dock, 2009.

A tribute to the fishermen and their families. These two sculptures – the little girl has had a bouquet of flowers added – stand at the entrance to the river, waving after having seen their husband/father's boat coming home, as had generations of spouses and their children. The relief on their faces is mixed with apprehension.

Yarmouth, three English towns renowned for their large wooden fishing smacks. These larger smacks, crewed by four men and a boy, trawled as far away as the north-west coast of Scotland with a 50ft-wide beam trawl on 150 fathoms of rope, staying out for almost a week. The smaller nobbies fished the bay, and as far south as Cardigan Bay, and as far north to the Clyde and Islay.

In the early 1890s the fleet reached its maximum of ninety-five sailing boats working from the port. In 1893 the 60ft *Harriet*, FD111, was launched for the Leadbetter family, a boat that fished from Fleetwood until about 1978 (the last wooden boat working from there), after which she was left to rot in the Duddon Estuary. She was subsequently turned into a sort of holiday hut before being acquired by the Fleetwood Museum where she remains today as a static exhibit, and where it is hoped she will ultimately be restored. Incidentally, the museum also has the nobby *Judy* in their collection. The otter trawl superseded the old heavy beam trawl in 1894.

The Wyre Dock had been completed in 1877, though the majority of the trade to begin with was commercial for the fishing fleet could land on Jubilee Quay free of charge. Though there had been attempts to use steam to propel fishing boats at Fleetwood, it was not until 1891 that the first steam trawler, the *Lark*, arrived. At 99ft long and grossing 133 tons, she was massive compared to the sailing smacks. Under her skipper Harry Bird, she was successful from the onset, much to the chagrin of the smacksmen who believed the noisy engine and black smoke would scare away the fish.

At the same time, with the opening of Preston Dock and, two years later, the Manchester Ship Canal, Fleetwood lost most of its commercial trade. The fishing industry was largely all that remained. The survival of this was boosted by the arrival of the so-called A.B.C. steam trawler fleet of Moody & Kelly (who had owned the *Lark*) which transferred from the east coast, the name A.B.C. coming from the fact that each boat was named alphabetically. Most of these iron vessels, built in Hull, had wet wells, enabling the catch to be kept alive until unloaded. However, they soon returned to fish the North Sea for a reason that is not clear.

About the same time the Hull firm of Kelsall Brothers & Beeching, one of the largest in the country, started to use Fleetwood, gaining exclusive use of the Jubilee Quay. For four years, up to 1897, they had as many as thirty-two trawlers working from the port. Then, just as the nineteenth century was about to close and Fleetwood was on the threshold of a depression, two trawlers of the James H. Marr fleet arrived from Hull. In 1902 James Marr & Sons (Fleetwood) Ltd was established to pursue the huge stocks of hake that existed on grounds north-west of Scotland, to the west of Ireland and off Rockall and the Faeroes. Fleetwood, from that time on, became renowned for its hake, though it must be added that it was a smacksman, Samuel Pearson Colley, who first successfully introduced hake into the Fleetwood market aboard the smack *Alice* in 1880.

Fleetwood's support industry for the fishermen and their boats grew rapidly in the early years of the twentieth century. As trips became longer there was a greater need for provisioning, re-bunkering and repairing. Dock Street became a mass of associated industries catering for all aspects of the fishing industry, from rope-makers to riveters, net-makers to engineers and coal yards to ice houses. Fish-liver oil, boiled and barrelled aboard, was landed for processing. All this onshore processing created many other jobs for the local population. By 1912 there were seventy-seven steam trawlers and thirty-four sailing smacks registered at the port, by which time there was a dedicated fishing dock, a station on the Wyre Dock and a brand new covered market.

Although some thirty steam trawlers were lost on active Admiralty service during the First World War, many returned to fish afterwards. However, sailing smacks, which had gained from the adoption of steam capstans in about 1900, continued to work, many having petrol/paraffin engines added after the 1920s. By 1933 there were 143 steam trawlers and many of these continued to work right through the Second World War, most commandeered by the Admiralty, while only two continued fishing from Fleetwood supplying badly needed food. When fishing resumed full-time after the war, the newer trawlers were up to 170ft in length and were staying at sea for three weeks at a time, with echo-sounders and radios fitted.

In the 1950s diesel engines replaced the coal-fired engines in the steam trawlers, though in the next decade a new purpose-built fleet of deep-sea motor trawlers developed and quickly grew. These, as before, fished as far away as Iceland, Bear Island and the White Sea with a crew of some fifteen. The last coal-fired steam trawler in fact to work from Fleetwood was the *Lord Lloyd* which was scrapped in 1963.

The diesel-engine vessels were, at first, side trawlers until stern trawlers arrived in 1966, the first being Marr's *Criscilla*, FD261. This vessel had improved accommodation aboard and a shelter-deck, which greatly improved the life of the men aboard. Another innovation, perhaps more important to her profitability, was her freezing ability, whereby the fish could be frozen into blocks whilst at sea. However, with Iceland extending its territorial limits in the 1970s and the subsequent Cod Wars, Fleetwood's days were numbered.

At the same time Britain entered the Common Market and threw away its fishing grounds to the rest of Europe. In 1970 there were over 100 vessels in the port but by 1984 this had drastically declined to two of the former trawlers, though there were still just over forty inshore boats. Today only the few inshore boats mentioned operate from Fleetwood as the fishing industry returns to its level of 150 years ago.

The wheel seems to have turned a full circle. All people have left are the memories, some good and some bad. Some of the old fishermen have their stories to tell – as did Frank Clarkson and Dick Massey briefly to me – whilst others consider the likes of the Marr family as part of the problems in today's fishing, where the powerful are able to lobby government and the small folk are forced out. But as Fleetwood, with its two fine stone lighthouses and the sentimental sculptures and outlet shopping, staggers on, one can only wonder what Sir Peter Hesketh would think.

TWO

LIFE ON THE LOWSHORE OF THE RIVER LUNE

Unlike Fleetwood, Lancaster has been the principal town and port of North West England since before Roman times, when it was the crossing point of the river Lune that flows down from the far eastern fells of the Lake District and emerges into Morecambe Bay at Sunderland Point. However, with the ever increasing size of ships, the river was eventually too shallow to accommodate these larger vessels by the beginning of the eighteenth century. Thus Sunderland Point became the point of discharge of cargoes, which were then transferred to smaller vessels for the onward few miles upstream.

By about 1715 local merchant Robert Lawson had built warehouses and other buildings alongside the existing stone jetty to supply the ships. Some goods were also distributed locally by being placed onto horse and carts. Later on, towards the end of that century, the Lancaster Canal was built to link the limestone deposits and gunpowder mills of Kendal with the coal fields of Wigan to the south. Limestone could then be burnt to fertilise the fields of Lancashire, whilst coal could be delivered to Preston, Lancaster and Kendal as well places in between.

Whilst plans were being laid out for the canal, Glasson Dock, built on the opposite river bank, had opened in 1787 and become the dropping off port for the Lancaster trade. The canal was fully operational according to 1819 plans, and seven years later a branch to Glasson was completed which linked the canal with the sea.

As well as moving goods around the hinterland and enabling rural development to gather speed, a fast packet service was also developed on the canal in the 1830s so that passengers could be sped along, pulled by galloping horses, between Preston and Lancaster. Travelling at 9mph, with the horses changed every 4 miles, passengers were served drink and food by a steward and the boats were warmed in winter. This really was first-class travel compared to a horse and carriage. However, with the arrival of the railway to Lancaster in 1846, the service soon ended. Train travel was both much quicker and more in vogue.

By the end of the eighteenth century Sunderland Point had become totally redundant and the area dubbed 'Cape Famine', an obvious reference to poverty amongst those who had previously worked to supply the shipping. Although plans had been made to build more facilities and to expand, nothing materialised. Much of the problem was that, in addition to the growth of Lancaster and the creation of Glasson Dock with its purpose-built facilities, twice a day the tide isolated the village. On top of that, the anchorage was exposed to the open sea. In the fast developing world Sunderland Point did not stand a chance.

Glasson Dock thrived as goods were moved through it. The shipyard of Nicholson & Sons built some fifty ships and later Nicholson & Marsh, as it became, built Morecambe Bay nobbies for shrimp fishing and whammel boats for the salmon fishery. The railway arrived in Glasson in 1883 and led to the complete demise of the Canal Company two years later, though some coal barges continued to serve the Lancashire cotton mills to the south. Today much of the canal remains usable for pleasure purposes and Glasson Dock continues to import timber and fertiliser.

Sunderland Point became a popular sea-bathing resort in the early part of the nineteenth century, and a regatta was held from the 1820s until Morecambe overshadowed it. More recently, in the first part of the twentieth century, it became a popular place for quiet holidays with locals renting out their houses to visitors in the season. Families came year after year for the scenery, bird watching and solitude, though after the Second World War the trend declined and the owners began to take up permanent residence once again. Today a third of the properties are privately owned whilst the rest belong to the Gilchrist Estate and are rented out.

Fishing is one occupation that has not completely declined at Sunderland Point over the generations, and a few fishermen also acted as pilots to incoming vessels. Most fishermen also worked the land and today few remain, though one of the last is Tom Smith, now in his seventies but still working. He is the third generation of Smiths to have lived at the river edge and to have worked on the riverbank, and sadly he could well be the last. It was his grandfather who moved across from Glasson Dock in 1901 and eked out a living by fishing and farming. Fishing was mostly for salmon and mussels which were plentiful back then.

Tom's father was almost exclusively a fisherman and Tom, who was born in the 1930s, started fishing with him in 1948. As well as the salmon and mussels, they set stake nets of 200 yards long on 4ft stakes that all had to be wriggled into the sand. Once the tide ebbed they fished out the flounders, though often they had to remove all sorts of rubbish that the river had brought down at the same time. Tom remembers the hedge baulk at Plover Scar, over the other side of the river by the lighthouse. It was just below the lighthouse and he recalls it was still in operation in the 1960s.

I visited Tom, and sitting in his bright kitchen, drinking coffee and eating wonderful homemade cakes, he told me a bit about his life. I started by asking about the flukes:

The 1912 Glasson Dock Regatta where swimmers dive off a pontoon into the canal basin.

Two years later, the 1914 Glasson Dock Regatta had adopted a local nobby as a diving platform, though the swimmers still dived into the canal basin which must have been severely polluted.

A fine evening view from Sunderland Point, with a steam tug towing two sailing vessels down river. Three nobbies are anchored off as is one whammel boat.

This view of a trading smack is said to be at Sunderland Point *c.*1880, though the position of the photo is hard to discern. The double-ended fishing boat seems somewhat out of place too, having possibly come from the Isle of Man.

This is the approach road to the Point in the early 1920s. The nobby drawn up the beach is the *Peggy*, belonging to the Gardner family.

I think the only place you'll find flounder fishing now will be across Flookburgh when it's in season. Rick Stein didn't do us any harm, he came up with a programme a few years ago, he was across at Flookburgh, he dressed them up with all sorts of nonsensical sauces but we like them as there are. Filleted of course, they're lovely. You can grill them if you want to but we fry them, in oil, bit of flour on to stop them spitting and cracking. I bring them in maybe this evening and tomorrow they're filleted and frozen. We've eaten them in their second season, the fish are prefect. I've had twenty packs to see us through the winter. Several years back, they weren't here, they were not in the river, they were pushed out with flood water. As an example of that, somebody was trawling for flat fish, a chap called Couch I think, he had a boat called *Judy*, one of the big prawners, bigger than the Crossfield nobbies that were built for Morecambe. It was a fine boat, a big one, a big prawner, had been a sailing vessel, I would think, he was one of the last from Jubilee Quay at Fleetwood. He knew every inch because I saw him come up here and he hauled down the channel here and he let go again and trawled up beyond Glasson Dock so he must have known it because there's training walls and stuff about, and it was a time when there was a lot of flood water out and I understand he had a bag of 120 stone of flounder.

When the mussels took off again, they were musselling when there's an 'R' in the month, like oysters and cockles, September to April, was my father's cockle season, it fitted in perfect. There's been time when they've been permitted to take mussels completely out of season. They must have been shot. But cockles, there was an old rhyme which said 'cockles and ray were best in May', so there's a different approach

Working the hedge baulk at Plover Scar, the lighthouse is just visible above the weir.

Another view of the Plover Scar weir, with several men looking on as the weir is emptied with small hand nets. The construction of the weir is easily seen. (Photo courtesy of Lancaster Maritime Museum)

Fisherman William Townley at Sunderland Point in 1932. He carries two tiernals which were used to carry all sorts of fish and shellfish.

Three fishermen outside their fishing shed at Bazil Point alongside the river Lune.

to cockles as though spring was not the best time. But I think cockles were better in the autumn, back end of the year. My father would start musselling in September, lay them in beds on the foreshore, just on the skeers, protect them with a coping of stones, put a scarecrow or two to keep the oyster catchers from a free feed, and then in October his orders would start coming in for bait mussels for the east coast. They used to ship them by railway from Glasson Dock to Filey and Scarborough. All the fishermen locally had orders because these mussels were splendid bait for their long lines. He worked for years with the same families, it was a good rapport, trust between fishermen, even the empty bags were bundled up and sent back by the rail which came to Glasson Dock. And they were doing it when I started work. With the *Mary*, the whammel boat, I've sailed to Glasson Dock with a ton of mussels in that boat, twenty bags to offload, then it fell through on their side and we started musselling for food. And they went to Lytham, but there was a little bit of rogue in them, sometimes they didn't pick up when they should.

Tom has in fact four boats: two whammel-net boats, *Sirius*, LR33, built by Jack Woodhouse of Overton for James 'Shirley' Gardner and Tommy Spencer in 1923; *Mary*, LR53, also built by Jack Woodhouse for Tom Smith's father in 1937 and named after his mother; *Moss Rose*, LR90, built in 1922 and *William Arnold*, LR173, a fibreglass whammel boat.

Jack Woodhouse had a yard in the middle of Overton and built a variety of vessels including whammel boats, prawners, mussel boats and pleasure craft for

Three salmon fishermen posing for the camera whilst they wait for the tide, perhaps taken by one of the holiday-makers. Note their ganseys, the uniform of fishermen throughout the country.

Most of the population of Sunderland Point were employed in either pilotage or fishing. Here several sea salts pose with some of their families.

Morecambe. It is said that the company was first started in 1663 and they ceased building in 1914. The fibreglass whammel boat was built by Bill Bayliss when he was building these in Overton. I counted seven such fibreglass boats on the beach at Sunderland Point. The original wooden boat, built in Sandside near Whitby, from which the mould was taken, still lies in the marsh above the beach.

Tom continued:

> We put a new stern and new gunwales and decking into *Moss Rose*. She's a Morecambe pleasure boat which we had converted as a family, my brother and father and I, forty, fifty years ago now nearly. She's on the marsh a lot, I had a new diesel engine put in it when I took possession of her completely. £175 we paid for the boat, just before she was put on the canal. That would have been the end of her, wouldn't it, pretty quickly.

She was the first Morecambe pleasure boat to be sold off because she wasn't big enough to run two of her crew when they had to have two men aboard for anymore over twelve passengers. So I got a new Petter diesel in it, hand start, 15hp only, there yet and running, only had it going last week. Never had a new injector in it, twin cylinder, grand job, made for people like me who didn't know a thing about engines. Tripping was good for the fishermen. My father used to do it in the *Mary*, charged two shillings for ten or fifteen people, round the lighthouse, up to Glasson Dock and back, depending on the weather. That was good money when most folk round here earned a pound a week.

I asked him about shrimping, whether he had fished it:

Father never fished with a horse, didn't want the trouble. I bought a horse as soon as I could afford one and I was the last horse and cart fisherman here shrimping. I didn't do the cockles but it was very successful shrimping, from Heysham Harbour, what we call Heysham Lake, over the sands, going up towards Middleton, and then out. Just water, horse and weights and going in the dark, very often ankle deep. The finest shrimps ever seen in Morecambe Bay down there over the years. Absolutely superb they were. We used a 14ft trawl. I didn't use the shanks like they do over the bay with the tractors. I used a straight trawl that could transfer to a small boat. Normal beam trawl, fastened to the back of the cart, on a bit of a run of rope so it could adjust itself and we trundled into the water. I had a super little horse, he was superb, lead me a dance sometimes when we were away, not when he was running free on the marsh. But he'd tackle anything. And it was only an autumn job, so it was September right through to about Christmas, so we were going in the frost. I took him when it wasn't fit out and he always tackled it. But they weren't always there. There were bad seasons as well as good. I've seen ten minutes fill a trawl up. I once dropped into a gulley that ran out over the sands and it had created a little bay, and I thought it quite firm, there'd been no great deluge of rain or anything, and I trawled across the rough pool, it's called, well up towards Heysham, and I said 'what's to do old lad you're full', and just that I had a fish box on board, like a flat cart, rather like the rag and bone collector would use, you know, in the old days, and I filled a six stone fish box. The net was behind the cart and the cart had all your gear on, fish hamper. You pulled to the side. Your horse stood just out of water and you tried to catch them before they came out of water and get the sand out of them. And then you'd have a fish hamper on board to tip them in and then walk back in the water to swill all the sand out of them. Empty through the cod end. Like with the boats. When I first started there was some lovely plaice and turbot down there, you could pick them up with a horse. They were so shallow. I mean you went in up to 2, 3 or 4ft sometimes, but they were sometimes so far inshore that you went in the daytime late in the season in November you couldn't find them with the boats, you couldn't get onto them, but you went in the evening tide with the horse and they were as thick as pitch, they were absolutely solid. They were there somewhere and you couldn't get close enough for them. I've seen it with the iron on the side, trawling on a

Lug-rigged whammel boats at Sunderland Point. Most of these boats were built at nearby Overton until the second half of the twentieth century when fibreglass whammel boats were introduced – again from Overton, though from a different boatbuilder.

little bank side, the one trawl iron, so a beam that high from the irons actually showing above the surface, and the shrimps were in that. You could feel around your boots, they were brilliant, my father had never seen such quality. You'd bring them home with a box lid on or a wet bag on, take them into the building with the light on because you brought them home to boil as against boiling aboard the boats. You were only away four hours so everything was fresh. And you'd take the lid off the box under the electric light and they'd be all over the floor, they'd be changing colour, taking on a paler hue under the light while they crawled away across the concrete. Like little monsters they were. They're beautiful they were. I liked working nights, digging under a full moon. It's sad it's declined so rapidly.

A lone fisherman in his clinker-built whammel boat.

What made you stop?

Well, he died, the horse that is, he was thirty-three, I had him twenty-nine years. He was only four when he came, but he was wise little bird. Little black and white fourteen and a half hands, like a Dale cob. Very robust, very strongly built, fit as a fiddle he was till he went. I did get another horse put into my care, a lovely brown mare. I went with her once or twice, she was quite capable but they don't all want water, or the soft sand; you got to get one that is amenable to, and she wasn't.

I would like to get back to it, if only to show Thomas [his son] what it's about. But with a boat of course, with a boiler on board. I have a fire boiler for *Moss Rose*, I've still got the old wood coal fire, I have one boiler left, still in condition that would go aboard.

You lash them into place of course, put a bit of asbestos underneath and round the back, I've no insurance of any sort anyway, on *Moss Rose*, and when you've got rigged up, you've got smoke and it looks like something out of the '*African Queen*', on that film with Bogart. She's 24ft and very stable so can carry the weight, as long as the bottom's sound. With the *William Arnold* you'd have a gas bottle in there, a gas burner, you got to mind what you're doing with them of course. You don't light them with a match anyway, you light them with a newspaper taper at a distance. There's many a man who's had his hair burnt off his arm with his shrimp boiler. Yes, I'd like to go shrimping again but let me get *Mary* rigged up first though I've not had time. Mother died before we could relaunch her. She saw her made whole again and even said it would be nice to have a sail in *Mary* again which she would do with my father. She was the last whammel boat built by Woodhouse.

Whammel-netting for salmon is drift-netting and the name seems unique to the river Lune. Elsewhere they just call it drift-netting. Using the boat, the net is payed out into the river downstream as the tide ebbs. It can only be used in the lower reaches of the river and the fish become entrapped by their gills. It is a relatively new technique on Lancashire's rivers, introduced around the end of the nineteenth century.

Today there are seven whammel licences, twelve for the haff-net men and one for a seine-net, the latter locally called a draw-net. This involves setting a net out from the riverbank by boat, around in a circle and back to the bank, before the whole net is hauled into the bank. The fishery is leased by a farm family, not looked on from a fishing point of view, from just across the river, the farm on the other side, Crook Farm, quite a successful farm, according to Tom.

On the other hand, haff-netting is an ancient practice, perhaps dating back to the tenth century. On the low ebb and early flood, the netsman stands in the river facing the current with his 18ft-long haff-net. When a salmon strikes the net he quickly raises the frame up and flips it over to trap the fish, which is then killed with a good thump with his 'priest'.

Tom continued:

The licence is £400 a year now and they've put on another £100 odd over the last few years. Over the same period we've had some of the most contrasting years in my lifetime, 2003 you couldn't work in an open boat in daytime. The heat was tremendous for me and the garden went to waste. And 2004 it went the other way and 2005 was cold. And again in 2007 and 2008. Never known such a cold July as this last one, it was nearly freezing. I think August was the only month in the year when there wasn't frost in the UK. There was probably frost in Scotland on the high ground. Then the flood water afterwards. This year we had the heat to start with and now the rain. I used *Sirius*, if it was convenient. But, so often, especially when you've had a massive amount of floodwater, the channel is so shallow, it's taken so much sand off and put it in the bottom, that I have a 12ft punt I borrow from a friend, one of Professor Bill Bayliss's, a really handy, very buoyant, boat. Two of us and 300 yards of netting, it's like a cork, it's super, and of

Jack Woodhouse's boatbuilding yard at Overton. Here a whammel boat is being completed.
(Photo courtesy of Lancaster Maritime Museum)

Woodhouse also built larger vessels. Here a nobby is about to have her deck fitted.

A fisherman demonstrating the haff-net to the camera at Bazil Point in the 1890s.

A push-net being operated in shallow water by fisherman James Rogerson of Morecambe in the river Keer.

Haff-net fishermen in the river Lune near Ashton Hall. The fishermen are (from left to right): William Alston, Richard Shaw, Jack Jackson and James Wilson.

Tom Smith standing alongside his whammel boat *Mary* in 2009.

Sirius, Tom's other traditional whammel boat, outside his storage shed at Sunderland Point.

course from the point of sculling across it's quite lightweight. So I've used that quite a lot, even to the bottom bar. About an hour and a half to row, with tide with you. And make sure you get back before it starts kicking up a stink. 2002 was some of the best fishing I've ever had in my life for salmon. I know I broke my own landing record and so it was a bit of a let down to have the weather turn against us for several of years. And normally I should have made it up with sprat fishing, whitebait. There's more shrimps boats rigged up here for part-time than there's been for ten years and they couldn't get a trip in.

What about the whitebait?

We use moor nets for it which is the beauty of it. You're not sat out. You need to fish them, either drop onto them with a whammel boat or a punt and pick them out of the water or if things are a bit meagre you can wade in and go with a wheelbarrow and have a hamper with you on the water's edge, just off the roadside. Autumn's probably one of the most productive times but any time through winter but whitebait we'll leave those in the chill, if there's too much ice or chilled flood water but you'll get sprats which again is a very limited market. I mean we used to sell thousands of pounds but it was to a dealer who had a zoo contract, a really great chap called Ross somebody from Fleetwood. But we lost the market when he packed up and no one else seemed to get a grip on it. Now, one of the reasons that whitebait, which was probably the most remunerative fishery I've ever followed, went down hill was they started to bring

Tom Smith's fibreglass whammel boat *William Arnold*, LR 173, built in Overton by Bill Bayliss.

in what they call free flow IQF whitebait from abroad – individually quick frozen – because the chefs preferred something like a bag of chips, they could have a helping instead of thawing a whole pack out. The Morecambe men actually went to Fleetwood to try and blast freeze their IQF, but our whitebait, true whitebait, was burnt. It won't stand it, it didn't work. They're far better frozen in their own clean liquid, in half pound or pound packs, whatever you want. They had a good shelf life and could be good in two years' time, frozen like that. It's an oily fish. When my brother was alive, a fisherman professionally, a character known far and wide and was quite a one for a dare, we sent some to Birmingham. In fact, we used to go when there was a passenger service from Morecambe promenade, in cold winter nights, and put frozen boxes and send them to Grimsby and they got a good reception there. By passenger train. There was a boat service to London from Heysham. But we sent some to Birmingham, can't remember the company at all, but at the time there was 70 tons frozen down of whitebait and they were offering about 25p a pound, a long while ago, want about two quid now, would be moderate wouldn't it if you think how much they are a portion, and they sent us 27p because of the quality, they volunteered a higher price. That really clinched it that we could compete. And then, the IQF came in and they weren't real whitebait, they were bigger, they had to be bigger, they were the size of half sprats as we call them. And in fact the chap I deal with now at Fleetwood has bemoaned the fact, he says I know someone who can get real whitebait, he's tried and tried them, they're a load of rubbish so I really want to try and get back into it. With people going

A nice view of a whammel boat at anchor looking across the river Lune to the marina at Glasson Dock.

Two fishing boats of uncertain vintage and build at Glasson Dock in about 1995. Judging by the gear aboard, both had been shrimping at the time.

abroad there's a much greater potential for door sales for things like that. They tried things and I'm prepared to freeze down to quarter pounds, you know, 120 grams – I've had to go metric of course – soothe the weights and measures people if they happen to come incognito. I don't know because everyone converts back. I believe for ten years you can keep the imperial measure on hand, but you must advertise in metric. But people come in and say I'll have a pound of that. I'll put some of them in a bag. In fact I was very relieved sometime in autumn a chap came in, broad Irish accent, he said 'would you be selling me a stone of 'tatoes'. I said thank God someone speaks English.

So whilst the fishing falls on hard times, Tom looks after his market garden with its potatoes, beans and fruit, an extensive area that the youngest amongst us would labour hard to have looking so healthy. Come August and hopefully the salmon will run so he can go whammeling. In the meantime, the spars and sails for his two wooden whammel boats await in his shed, the boats get fettled and the other nets ready to fish.

This life of his, so reminiscent of an older age, seems so detached in today's society – and indeed would be for him if his wife didn't work – but that fellows like Tom Smith still work the twin lives of fisher and farmer serve as a memory that, in a bygone age, most of the country dwellers lived that way. For the inhabitants around Morecambe Bay this was the norm back in the nineteenth century and before, until the railways opened up the area. To them it was both necessity and tradition, ideal by the standards of many today, but overall really an arduous life, out in all weathers, and certainly with no luxuries of travel.

THREE

HEYSHAM –
A MISSED OPPORTUNITY
OR A RED HERRING?

Around the tip of Sunderland Point and across the shoulder of the Middleton Sands lies the modern harbour of Heysham, along with its unsightly and stark nuclear power stations and beyond the serenity of Heysham Head.

Tucked neatly beneath the Head, Lower Heysham today is still a sleepy little village reminiscent of that quintessential English hamlet associated with more traditional southern parts of the country. Walking down the narrow main street in the bright afternoon sunshine, the flower tubs and shrubs alight with colour and scent, there's an air of tranquillity quite unlike elsewhere around Morecambe Bay. The old village water pump lies hidden within its stone frame and 'Cockle Cottage' is the only reminder of an occupation now largely forgotten. Tea rooms sell nettle tea, a local speciality, a brew that seems to have stemmed from Granny Hutchinson's nettle drink, or beer as some call it. It is as if things haven't changed much around here.

The slipway onto the beach is a bit different and more abrupt, with its large white-painted 'Keep off' and 'Private' signs around. A tractor is being painted in red oxide and alongside is what appears to be a similarly painted trolley, with trawl-nets being made up by a fellow. We talk briefly of shrimps for the owner is hoping to start out on the sands in a week's time. 'How far out d'you go?' I ask. 'Not far, just down to the low water mark, this side of that boat.' He points to a boat anchored of the beach, maybe 800 yards away. 'Not eight miles then!' I reply. He laughs at the thought of it. Back in the Heritage Centre I ask how many families fish. Seems it used to be the Baxters and Braids but they no longer do, I'm told. These days folk go out at weekends when they aren't working. Shrimping has become a sort of playtime activity here, a way to supplement an income and have a good feed at the same time. No wonder he laughed!

The point at Heysham where the lane leads down to the beach, the main access for the fishermen to the sands.

The shore road again, some time later with Charlie Edmondson's jewellery and seashell stall and Miss Schofield's ice-cream and sweets stall, both catering for the 'select' visitors Heysham attracted.

Victorian holiday-makers in resplendent beachwear.

The harbour at Heysham showing the North Quay that was built to attract fishing boats, though never did.

Today the shore road still ends up on the beach. Here the shrimp fishermen prepare their cart.

Above: The colourful frontage of 'Cockle Cottage' seems to be fairly typical of Heysham's attractive main street in the summer of 2009.

Left: Nettle beer has always been a delicacy from Heysham.

But wasn't it always like that in a way? Not playtime for sure, but subsistence fishing. The combination of working the land and gathering in from the sea, here, as elsewhere around the coast. Mussels and cockles collected from the skeers and carried ashore by horse and cart, fish collected from baulks and stake-nets. Heysham might have stayed that way for generations but for developments very late in the nineteenth century that soon disturbed the peace and brought about monumental change.

The Midland Railway Company decided that to create a link across to Ireland a deep-water harbour had to be built. And Heysham was their favoured spot. Although nearby Morecambe had been the stopping off point for sailing boats over to Dublin for many years, that service finished in August 1904 when the new port opened, seven years after construction work had commenced. Belfast was later added as a destination but today it is principally the port for the Isle of Man. Part of the original plan was the building of a fish quay on its northern side in the hope that fishing boats from Morecambe and the surrounding areas would come in to land but that aspiration never materialised.

However, from the point of view of the pretty part of the village, the port is over the headland, far enough away to be ignored. In between, Britain's first holiday camp at Heysham Towers once amused visitors on the headland, brought in from all over. Heysham, it seems, was more concerned with tourism, and when, in the 1930s, sewerage caused the mussels in the bay to be declared unfit for human consumption, those that still fished had even less incentive to venture out onto the sands. Since then it seems Heysham has never really bothered to benefit from the diversity of fish available just off its shores.

FOUR

MORECAMBE – THE OLD FISHING COMMUNITY OF POULTON-LE-SANDS

I found it almost impossible to work out where Heysham ends and Morecambe starts. Perhaps it is the point at which the seafront starts, a mass of thousands of huge stone boulders strategically placed as part of the flood defences from the sea which has, in the past, caused havoc and destruction to the seaside town. There were once two grand piers, both of which met sticky ends. The earlier, 1869-built, pier caught fire in 1933, whilst the West End Pier, built in 1896, almost broke in half during a severe storm in 1927. It was not until 1968 that it was ultimately washed away by the notoriously violent storm of that year. However, with regard to these severe weather bouts, the 1907 storm seems to have been the one that went down in history as resulting in the most damage.

Talking of breaking, the old stone pier was once the home of T. W. Ward & Co., a company specialising in the breaking up of ships in what was once described as Morecambe Harbour, the space between the first mentioned pier and an earlier built wooden landing stage that was used for embarking passengers aboard tripper boats. Between 1905 and 1933 ships were scrapped, an activity which many thought an intrusion into the gentle seaside town life. Surprisingly, though, many of the visitors disagreed and paid a fee to enter the yard to view the breaking process. Probably this made quite a contrast to the everyday beach activities, beauty pageants and musical troupes – a bit of serious ship-breaking clamour. Once the pier had burnt, the Council decided on development and the breakers were told to go.

However, with the piers gone, and the landing stage removed in the 1950s, the fortunes of Morecambe declined as package holidays abroad grew, especially in the 1970s. Today a huge amount of investment has flowed into the town, and nowhere is this more obvious than upon the promenade with its various sculptures – cast-iron birds perched on bollards and others seemingly in flight on the face of fencing. Bird

An old print showing women collecting mussels.

At low tide the seagulls were the most voracious of feeders and the fisherman had to be quick to collect their catch.

Collecting the fish at low tide in the 1950s.

A lone fisherman setting his stake nets out on the sand, accompanied by his faithful hound.

Mussel fishermen at work. Over 100 fishermen were once employed in Morecambe in the mussel fishery alone.

When the mussels were collected by boat the catch was transferred to a cart for transporting ashore. After 1848 much was shipped away by train inland.

Often the mussels were deposited on the promenade until the council stopped the practice.

Washing a basket – or tiernal – of mussels at Morecambe. (Photo courtesy of Lancaster Maritime Museum)

life in the bay still attracts visitors but Morecambe is a very different place from its heyday when folk flocked to the Winter Gardens.

Why did they come? Not only because of the good railway link but because in 1909 its merits were said to be numerous: a mild temperature, a low death rate amongst its inhabitants, a noble bay, proximity to the Lake District, its bracing air that had a low bacteria count, a high oxygen yield and a general lack of fog – by which they probably meant smog.

But before all this, before 1889, what is now Morecambe was the three villages of Bare, Poulton-le-Sands and Torrisholme which were combined that year to form the single town which through much of the twentieth century was a renowned seaside destination for thousands of annual visitors. Almost a century earlier, well before the advent of the railway, and being the village closest to the beach, Poulton-le-Sands was itself a thriving place for holiday-makers from the nearby inland industrial towns. At the same time it had already had a fishing community that stretched back many generations.

Two of the first travellers in the nineteenth century to write of their experiences in Poulton were William Daniell and Richard Ayton, who were both journeying around Britain. Although Ayton was the chief writer whilst Daniell produced many fine aquatints of their travels, Daniell later undertook the journey on his own after the two parted in Kirkcudbright. They had begun their trip at Land's End in 1814 and arrived in Blackpool later that year, finding the place a popular bathing resort full of 'poor people from the manufacturing towns, who have a high opinion of the efficacy of bathing'.

After crossing Cocker Sands to Glasson Dock, they travelled to Lancaster before passing on to Poulton, being unable to find lodgings in Lancaster. Here, stopping at the first public house in the 'large village', they found it full up with sixty people. These people had come because it was the day of the first spring tide which appeared to draw them in for the 'partaking of the seawater'. At the time people were just beginning to recognise the beneficial health effects of bathing. Finding another public house where accommodation was available, they were horrified to find six people sleeping to a bed including many women who were 'so gaily attired'.

Seemingly, Poulton was brimming with visitors – six cartloads had arrived that day – and they weren't necessarily impressed with the goings-on between men and women! A shame really because the visitors were obviously enjoying themselves through imbibing and singing and other dubious activities! Though of course it was really the sea bathing they had come for. A few years later, in 1829, it is known that they were holding annual regattas consisting of a series of rowing matches, with the fishermen probably being instrumental in organising these events and possibly seizing the chance to show off their manliness to the female visitors!

However, Ayton and Daniell do not mention the fishing. The first mention of fishing, according to Kennerley, had already been made when the *Lancaster Gazette*

Here the mussels are being gathered from the shallow water over the skeers. Sometimes the profitable skeers were 2 miles offshore.

A good pile of mussels awaiting transport to market. (Photo courtesy of Lancaster Maritime Museum)

of 1803 mentioned '*Otter* fishing boat arrives with first catch of the season, 250 soles, plaice, cod etc.' The boat in question was of 34 tons – a fairly big boat by the standards of the day for fishing – and was built by John Brockbank at his local shipyard.

Fishing then, as we have seen was in most parts of the coast, was a subsistence occupation combined with farming the land. Commercial fishing had yet to arrive, though any surplus fish was always sold to make a few pennies, with the Poulton men selling theirs to Lancaster. In 1840, almost a generation later, a description of Poulton was given by Dr Edward de Vitre as part of an enquiry into sanitation. He found that the male population:

> …are almost exclusively engaged in fishing for herrings, flat-fish, shrimps, cockles and muscles [*sic*]. The nature of their occupation implies great exposure by day and night; and whilst they are so engaged, their families at home are picking and sorting produce of each previous catch…

The place obviously stank of discarded fish guts which were thrown outside their houses to rot as manure. The place was rife with poverty and the average life expectancy of a fisherman was only twenty-four years, partly due to disease and partly from the nature of their work. Many a boat was wrecked, as illustrated by the biggest

Bagging mussels at Kilnbrow, near Morecambe, in the 1930s.

A mussel boat unloading under the shadow of Morecambe's pier.

Mussels being riddled and bagged on the promenade at Morecambe, alongside the Calton Landing Stage Company's hut at the top of Lord Street, Morecambe.

accident when, in 1895, fifteen Morecambe boats were caught in an unexpected and sudden gale and four fishermen were drowned.

Though there were some 170 fishermen and their wives in 1861 according to the census, and 203 a decade later, there were boys who were also classed as fishermen, some of these lads being under fifteen years old. In 1871, for instance, there were twenty-five boys older than fifteen years and seven younger. Few old men fished, probably because they were dead if the average life expectancy was twenty-four, 'having cast off their moorings and sailed on the ebb for the rendezvous with Old Nick', as John Dyson puts it in *Business in Great Waters*.

Many believed they returned as seagulls, which might account for the vociferous and multitudinous nature of these sea birds. Maybe they needed some respite and freedom, for the lives of these fisherfolk are, with their daily hardships such as lack of fish, mending of torn nets and other endless tasks, hard indeed to imagine these days. Many of today's fishermen, with their electronic screens, automated nets and relatively homely accommodation are a poor reflection of their forbearers. Not that that is a bad thing, it just means that on the whole fishermen today do not rely upon the same skills to search out and land the fish. Nostalgia is a great thing unless it is you doing the job!

At the end of the nineteenth century, local fisherman Jack Mount started fishing at the age of ten. Going to bed at 7 p.m., he would be up at 1 a.m. Wearing oilskins, rubber boots, a woollen jersey and a sou'wester he went prawning with his dad, returning with 'three apple barrels full'. Then he had to get the salt and coal ready, presumably to boil them, and he earned himself 6*d* a week. Jack got his own boat, the *King Fisher*, when he was sixteen at a cost of £62. His great grandfather had been drowned whilst tending to his stake nets and got caught out by the flooding tide.

Although fishing had been practised at Poulton for generations, probably from time immemorial, one of the methods that the area was renowned for was a particular type of fish weir which was known locally as a 'hedge baulk'. It is believed that these hedge baulks were extremely old and it has been suggested that the monks of Furness Abbey owned the rights. Before that, further suggestions have been made that it was the Celts who first devised them, before the Romans came and encouraged their use. More recently, ownership has been traced back to the early eighteenth century, when at least one baulk was owned by a family with no other connection to fishing. Presumably this was simply rented out to fishermen.

All along the coast, from Heysham to the shore off Bare, these baulks have been sited for years and they consisted of an elaborate construction of stone, posts and hedging and their construction involves a great deal of skilful labour and expense. One such structure was sited on the largest skeer off Poulton, which consisted of five hedge baulks next to each other in a zigzag way. The names were often localised according to past fishers, such as Jacky John Skeer, and again one of the baulks on Old Skeer was called Old Dick Bond's.

To build a hedge baulk, the basic shape resembles a large 'V', with a wide open mouth which is laid out on a shelving beach. Stakes of oak – up to 12ft in length –

A typical mussel boat with lug sail, as built at Overton by Jack Woodhouse. (Photo courtesy of Lancaster Maritime Museum)

A collection of tripping boats. Many of these would be mussel boats that were adapted for the purpose during the mussel summer off-season.

More tripping boats. Sometimes nobbies would also undertake trips as far as Grange-over-Sands. (Photo courtesy of Lancaster Maritime Museum)

are driven into the sand using a heavy wooden mallet so that some 8ft of oak sticks out above the sand. It is constructed at right-angles to the shore, and the length of the arms of the 'V' are different, one being some 340 yards long whilst the other is some 270 yards, and the mouth of the baulk is 150 yards across. Between the oak posts are wattle hedging of hazel against which the sand will eventually silt up on the outside.

At the apex of the baulk is an intricate cage made out of netting over the posts, including over the top, where the fish end up. This netting has a stone foundation in which there is a grating to let fish out. Around this cage there is another low hedge along which water will flow out of the main baulk. There is also a sprat pool where small fish collect and a 'puzzle garden' which, to me, is indeed puzzling. Of all the various fish weirs and structures around Britain these surely are the most complicated and thorough in design. They were, however, deadly in their catching of fish, and 60,000 herrings in one go was not uncommon.

Morecambe Bay also has an assortment of other stake nets. Firstly, there is another type of baulk-net found all over the bay, from Glasson to Barrow, as well as into the Duddon Estuary to the north. Again this is a series of posts with nets hanging off, but an ingenious smaller baulk of timber – a much thinner piece of wood than the post – is tied to the top and bottom of the net and to the post at the top. When the tide floods in, the baulk lifts the bottom of the net to allow the fish to swim in whilst on the ebb the baulk sinks and closes the net. Other forms of 'fixed engines', as these structures are called, are the paddle-net, roa-net, poke-net, teedle-net and bag-net,

Above: Here a nobby has come alongside the landing stage to pick up trippers from a crowded promenade.

Opposite: The majority of these boats in this photograph appear to be purpose-built tripper boats with plenty of seating.

Two nobbies disembarking trippers on the landing stage alongside the pier.

Nobbies anchored off the stone quay at Morecambe, with ice floes on the sea. Possibly the cold winter of 1963.

Today, in 2009, one lone nobby, *Linda*, was anchored off the same stone quay.

'The bird chasing the golden fish' sculptures on the fencing by the stone quay.

all of which are similar in that they consist of suspended netting though their setting and method of action are very different.

These structures trapped all manner of fish but especially herring, codling, plaice, flounder and whitebait. Herring was plentiful and were available at a penny for as many as you could carry. Herring was also caught in drift-nets set from a boat in the autumn and early winter, and fishermen sailed as far north as Maryport. Herring though is a fickle fish and it might come to the bay year after year, and then suddenly disappear and not return for a number of years. Some years were good for the Poulton men and others disastrous.

1842, for example, was a bad year whilst on one night in January 1853 the boats were catching 600–1,000 herring per boat, which was described at the time as a 'fair catch'. In the 1870s they were catching more 'big fish' than shrimps, these fish being 'plaice, flounders and flukes', which suggests the herring take was poor. The differentiation between 'flounder' and 'fluke' is puzzling, as they are the same fish! By the 1920s the Poulton herring fishery was all but ended, though some continued being landed into Maryport in the late 1940s.

Shrimps were a major source of income as were cockles, and both will be discussed in later chapters. Pink shrimps – what they call sprawns in Poulton – were caught up to about Christmas. Musselling – the gathering of mussels (*Mytilus edulis*) – was a fishery that, like cockling, involved the whole family. In the early years of the twentieth century there were 100 Poulton men engaged in the mussel fishery.

There were two ways of collecting the mussels off the skeers: using a mussel craam from a boat or picking direct from the skeer at low water. The craam used in the mussel fishing was an iron rake with long and narrowly-spaced prongs on a long 12ft wooden pole, with which the mussels could be dislodged and brought up from the seabed into the boat. The method of collecting at low water had the advantage that only the biggest and best mussels could be picked. By-laws eventually said that any sold should exceed 2in in length. The season was, as everywhere, from September through to April so that, as Tom Smith informed us, the months without an 'R' in them (May to August) were supposedly closed. Some, of course, did not use boats but merely set out with a horse and cart, a short-handled craam and a 'tiernal', a special basket with a hooped handle for carrying the mussels in.

The Poulton men had gathered mussels from Heysham skeers for generations, as had those from Heysham. However, the rights for this belonged to the Lords of the Manor of Heysham, though nothing was ever enforced. In 1873 the Lords suddenly decided to re-establish these rights by attempting to charge the mussellers a levy on every bag they removed. One fisherman, Robert Fawcett, continued fishing without making any payment and the Lords of the Manor took a lawsuit out against him for fishing without their permission.

The mussellers from both villages joined forces to support the fisherman and objected on the grounds that they had had access to this free fishery for many years. A court case followed in 1874 in which the jury agreed with the fishermen and they won their case. However, a case in 1802 had been brought against Christopher Orr,

The remains of a hedge baulk off the beach in 2009.

Three fishing boats at anchor off the stone quay in 2009. The outer two trawl for prawns during the summer.

a mussel fisherman, in which an award was made to Robert Bagot, as trustee to the Lords of the Manor, respecting the right to certain parts of the foreshore as belonging exclusively to Bagot.

Mussel boats were flat-bottomed craft that could sit on the skeers to be filled by hand. Most were clinker-built and many came from Crossfields of Arnside, of whom we will learn more in a subsequent chapter. They were originally about 15ft in length, but larger boats were built in the early nineteenth century and these were able to carry up to some twenty-five bags, each weighing one hundredweight. They were rigged with a small homemade sail – probably a lugsail or spritsail – and were also used for stowboating for sprats and whitebait and drift-netting for herring. The stow-net is a complicated system of netting set below the boat from its bow, into which the fish swim. In later years these mussel boats were built by Jack Woodhouse at his yard in Overton.

The mussel fishery was not huge. In 1890 some 2,360 tons of mussels were collected, though this had reduced to 823 tons eight years later. The mussels were brought ashore and riddled and cleaned, though the fishermen were persuaded to riddle them below the high-water mark and not on the Promenade in 1923.

By this time Morecambe was Morecambe, and Poulton just an area of the new town. The railway, which had reached Poulton in 1848 thanks to the North Western Railway Company which had chosen Poulton to develop their Ireland link, was used to transport the mussels to markets. Before that it was the canal taking them to

Lancaster and Preston. The very same railway had been bringing in visitors to the town since its arrival, bringing prosperity to all, including the fishermen who benefited from taking these visitors out for trips around the bay in summer.

However, as this prosperity grew so did Morecambe, while, at the same time, fishing declined. Most fishermen by then were members of the Morecambe Trawlers Co-operative, which was formed in 1919 to make the marketing of the fish and shellfish more reliable. Today the hedge baulks and other stake nets are gone – though signs of them are still visible on the skeers – and few boats come to Morecambe to land fish, a far cry from the 200 boats said to have been moored off Morecambe seafront at the peak of the shrimping era between 1900 and 1930.

Some fish is sold locally, both in shops and for visitors, and some goes out of the area, though the gross amount does not total that much. Sewerage problems led to a cessation of mussel fishing in the 1930s and again in the late twentieth century, and few are picked these days. Cocklers still gather on the Sands when the fishery is open, though few are born in the town. Fishing for Poulton-le-Sands is all but a distant memory in the sands of time in more ways than one.

FIVE

HEST BANK – COCKLING THE SANDS

From Silverdale to Kent sandside,
Where soil is sown with cockle shells.
(Old Flodden Ballard)

Today (summer 2009) the cockle fields of Morecambe Bay are quiet, fishing having been banned for the time being. However, these fishers of the bay had received more than their fair share of publicity when, on one cold night in February 2004, twenty-three Chinese cockle gatherers were drowned on the notoriously treacherous sands of the bay after failing to return from gathering. Some might have thought that cockling was a relatively new fishery. However, it is far from it, and cockles from Morecambe Bay have been gathered for generations.

In 1868–69 records show that £2,000 worth of mussels and cockles passed through Morecambe station. The occupation was primarily a family affair, and usually was the job of the women and children. Children were often reported absent from school during the season that was then mostly during winter and spring. Some say the women used to work barefoot throughout the year. Such a woman from Poulton-le-Sands was Mary Armistead who died there in 1826 at the age of ninety-three. She was known as Cockle Mary and is said to have walked to Lancaster each day with her cockles. She was reckoned to have walked the equivalent distance of three times around the world in doing so!

Although some used jumbos, the tamping tool of the cockler, at times, others looked for the tell-tale air bubbles in the sand that denote their presence. It was said that keen observers could spot 'the siphons of the cockle, the tubes through which the respiratory and food-containing currents of water are in- and exhaled'. Another trick was to tread the sands with the feet, thus bringing the cockles to the surface. It has been said that, at times when the use of jumbos was banned, children had planks of wood tied to their shoes to stamp around, unseen, to bring the cockles up.

The jumbo purportedly came about after the working mothers brought their babies out onto the sands in their cradles, the rockers of which were seen to bring the cockles to the surface as they rocked vigorously. According to A.M. Wakefield, who wrote an article in the *Pall Mall* magazine (vol.XVI, 1898), a woman with a two-week-old baby was heard to say she would not feel well again until she 'could get to t' cockles again'. However, seeing how they often started 'wading knee-deep', sometimes walking 10 miles out into the sands, and only allowed upon the cart when the channels they crossed were deep, it cannot have been a pleasant occupation by any description of the word.

Wet, cold and back-breaking, cockling was, at times of the year, the only source of income, and therefore a choice there was not. The channels themselves changed with the regularity of the moon, and an intense knowledge of the sands was indeed a necessity. Nevertheless, in the season there was a string of dots seen on the horizon, this being a line of carts, the figures working hard almost unseen. In general terms, though, cockling has not altered much in over a hundred years.

In those days the only mode of transport available to move the shellfish off the beach was by pony and cart, the pony, according to Wakefield again, looking 'as if it had been dried in the sand and salt water for centuries'. Much of what was picked was taken to the canalside at Bolton and shipped off to Preston. The same horses were also used to drag a shank-net over the sands to catch shrimps (more of that

A lone cockler comes ashore with his horse and cart at Hest Bank in the early part of the twentieth century.

A group of cocklers awaiting the tide at Silverdale. Cockling involved all members of the family, whatever the age. Here young boys and a girl join in with the older family members.

Each fishing family had their own horse and cart for working the sands. Their livelihood depended on the horse's welfare and so they were well tended.

The typical dress of a cockle woman.

Cocklers on the sands – they appear as dots on the horizon, bent over in their arduous work.

in another chapter). It was not until the early 1960s that tractors gradually replaced horsepower.

In the *Report of the Commission on Sea Fisheries 1879*, it was estimated that £5,000 worth of cockles were taken from the south side of Morecambe Bay and that it followed that some £20,000 worth were taken from the whole bay. Furthermore, they were unable to trace any decrease in the yield of the fishery. Thus it must have been a substantial fishery. In 1890 Wakefield tells us 3,162 tons were fished at Flookburgh, at an average price of £2 8s a ton. This was the best year ever known. Three years later it had shrunk to 1,335 tons. In 1895, after a devastating frost that killed much of the harvest, coupled with a poor demand, the take fell to 822 tons. But the next year it was down to a meagre 50 tons after a minimum size was introduced, although improvement followed, with the take increasing to 195 tons the following year. Such is the feisty nature of fishing! However, by 1911, 65,500 cwts (3,275 tons) were taken, indicating a substantial revival.

A few years ago I spent some time looking into the cockling of Britain in general, and more specifically that of the bay. One morning, a typically dirty sort of morning, saw me start the day in a car park on the east side of the bay. This was at what is locally known as Morecambe Lodge between Hest Bank and Bolton-le-Sands. It was easy to find – a yellow signpost said 'All Cockle Traffic', suggesting I was in the right place.

A few miles to the north were the notorious Warton Sands where the Chinese had perished. I had reckoned that, as four hours before low water was the time to head

Using the jumbo to bring the cockles to the surface. (Photo courtesy of Lancaster Maritime Museum)

Some cockle fishermen also set stake nets and would check these for any fish whilst en route to the cockle grounds.

Man and wife team with jumbo, horse and cart.

Here the horse is feeding from the up-turned cart whilst the owners go about the monotonous task of gathering the cockles.

A posed photograph of a man and wife team out on the sands.

out on to the sands, the first cocklers would appear about 10 a.m. At 9.30 a.m., just as I had brewed some fresh coffee, the first tractor towing a trailer rumbled down the track cut between the surrounding marshland. Walking down to the edge of the sands, mug in hand, I tapped on the driver's cab.

He turned out to be Paul, a Liverpool lad who was in charge of a gang of up to twenty-five pickers. 'Today probably eight or nine will turn up, it just depends on who wants to work,' he said. Most, it seemed, were either English, Welsh or Polish workers. After the disaster of the previous year there were no Chinese working at all. The Chinese had been working illegally and various official bodies such as the Health & Safety Executive, the police, local council, Immigration Authorities, Work & Pensions and North Western and North Wales Sea Fisheries Committee (NW&NWSFC), had since ensured that everyone was now legal. Since the end of 2003, permits with photos of the fishermen had been issued by the NW&NWSFC, which polices the inshore fishing sector up to 6 miles offshore between the Duddon Estuary and Cardigan. The disaster had made sure that the permit system was fully operational.

I asked Paul about the nature of the job:

We get about five hours out on the sands. Us locals, the British so to speak, go out four hours before low water but some of the others chance it and go earlier. But it can be pretty dangerous out there. We've lost three tractors before and that's £120,000 each. I took one up to the Solway a few weeks ago, but that's closed now.

Much of the catch was taken to the canal at nearby Bolton. Here the canal has been breached and drained for works to repair. The loss of the canal would have been a serious upset to many of the cockle families.

Signpost in 2005 denoting a substantial amount of cockle traffic.

I asked how much a man can earn:

> Price depends on time of year and quality but it's about £450 a ton right now. Most are going to South Wales just now where they are processed. Sometimes it's as high as £1,400. Each bloke picks by the bucket, two buckets to a bag and six bags is the average right now. Sometimes he can get a ton but the pickings aren't very good at the moment here. But they should recover in May and then the Spanish buyer comes and the price goes up. They should really close this area now.

When they were not picking there, it seems they could be at nearby Middleton, or on Pilling Sands, near Fleetwood, or Bardsea, across the bay, or Flookburgh, or the Solway, or further south on the Ribble Estuary, or even as far away as the Lafan Sands or Anglesey. But, as well as the Solway, both Bardsea and Lafan were closed at that time. That seemed to be the nature of the job, a juggle between different areas and a transient workforce. And the cold, back-breaking monotony of the reality of the job, of course.

On Paul's trailer were eight structures that consisted of a metal framework of two uprights and a crossbar, attached to a wooden base, about 4ft by 18in. These were the 'jumbos', or 'tamps' as many call them nowadays. All-wood versions have been used for possibly centuries to expose the cockles that lie about half an inch below the sand. They are rocked forwards and backwards to suck the cockles upwards. According to

The beach during the author's visit in 2005. The men had just arrived and were waiting for the tide to ebb.

Today's jumbos are made from steel, shown here with a group of Polish workers.

Kennerley, though, jumbos were not legal most of the year in the nineteenth century for it was believed they damaged the cockles. For a period in the twentieth century they were banned altogether.

As I talked to the affable Paul, a car arrived towing what turned out to be a mobile burger van. 'That's Pat and Jan,' Paul told me, 'they'll be busy soon. Pat's Irish and is in mourning after the Welsh defeated them.' he added, referring to a recent rugby match. I went over as soon as they were set up and commiserated with him on the Welsh rugby victory before buying a refill for my coffee mug and asking a few questions. 'Been here every day for over a year. We'll be busy for an hour or so, then they'll all go off and nothing will happen for four hours or so. Then we'll be busy again,' Pat told me in his Irish drawl. This really did seem to be a real enterprising business, selling bacon butties and burgers out on the sands.

As if to order, minibuses arrived in the car park, and vehicles began to rattle down the track and gather around us. Each towed trailers loaded with jumbos which emphasized the fact that they were certainly in use these days. Land Rovers, 4x4 vehicles, pick-ups, quad bikes, a couple of vans and two more tractors arrived, seemingly full of people. More pickers wandered down on foot.

As I drank my coffee I heard the talk of what sounded like Polish. I talked to a couple of the lads, one guy even being persuaded to demonstrate the use of a jumbo. Then the first of the convoy headed off, followed by a few more. Paul waited another half an hour before he too headed off with half a dozen Poles or more on his trailer. The rest of the vehicles followed suit and suddenly the beach was empty, save for Pat, Jan, the dog and me.

I sat on a tuft of the grassy marsh, watching through the binoculars at the figures already out on the sands. They were mustered in groups, seemingly floating on water, until the ebbing tide made the sands more obvious. I was reminded of the words of Rick Stein who had been here before and who wrote of 'looking across the shimmering haze of the flat white sand and blue horizon and seeing other high black shapes of tractors, like camels in the distance crossing the desert'.

The stick-like Lowry figures, some upright and some bent, silent and motionless from a distance of a mile or more, were busy filling their buckets. Using the jumbo to bring them to the surface, the cockles are raked using a craam, a three-pronged short fork, before they are sized. Local by-laws state that cockles that pass through the 20mm by 20mm mesh of a net-bag are undersized and therefore must be left on the sand. That's a regulation that started back in 1895 when, after a poor harvest, gauges were supplied by the Fishery Commissioners through which any cockles that fell through were to be left on the skeers. Interestingly both the craam and the jumbo have been referred to as 'engines', as many of the mechanical forms of fishing are referred to.

From my perch I wondered about the animosity I had been told about concerning the use of migrant labour. Far from being illegal immigrants, as the media spoke so disparagingly about, these were fully paid up members of the European Union and as such had the perfect right to work here. After all, they were doing a job that few locals wanted to do, and those that did, were participating. I could see no reason to doubt the validity of their presence.

Paul had told me that occasionally the police swoop, look at permits, sometimes booking vehicle drivers for such pathetic crimes as uneven tyre pressure. There was even one policeman assigned primarily to policing the cockle fishing. Presumably folk from the Inland Revenue and the Benefits Agency make their presence known to ensure all those involved are not moonlighting. With the bad press received nationally the previous year, there seemed no alternative but to be 100 per cent legal.

Bill Cook was, at the time, the senior scientist in charge of the NW&NWSFC and I found him in their offices housed in a wooden hut on the campus of Lancaster University. I began by asking him what he knew about the more recent history of cockles:

> In the harsh winter of 1962/63 the frost almost wiped out the entire stock which hardly survived until 1970. After that, in the 1970s and 1980s, it was a fishery that took control of itself although by-laws did prohibit the use of mechanised fishing, and controlled the minimum size.

In 1975, 1,680 hundredweights were landed. Then came the migrant gangs in 1987 after the failure of the Dutch fishery:

> There were big stocks here then, right the way up from the Dee up to Morecambe Bay. But it was an unlicensed and unregulated fishery then and there was no way to control the numbers of people fishing. Soon after, consent was given for the use of mechanical dredgers that were towed behind tractors, almost like mechanical harvesters. But this was stopped in 1991 because of the damage they were doing to the spat. Through the 1990s only a few hundred tons were taken each year but that all changed two years ago when a large boat arrived on the sands with hand pickers filling it up. 2001 was a good year with an abundance spread around the whole bay.

Although it was difficult to establish the exact amount taken, he reckoned DEFRA (Department of the Environment, Food and Rural Affairs) estimated that between 3,000 and 5,000 tons had been removed during each of the previous two years. Stocks were becoming low, and parts of the west of the bay had been closed.

We talked about their survival rates and the danger from the fishing, and echoing what I had previously read he said:

> Raking tends to destroy the spat so we prefer the jumbo. But cockles have a vast difference in recruitment from year to year. This depends on the weather, sea temperature and the predators, such as the seagull and the oyster catcher. In 2002 there was no settlement but a good spat followed in 2003. Growth is variable, but on average it takes one year to eighteen months for them to reach maturity.

Left: 'John' demonstrating the rocking back and forth motion of the jumbo that makes the cockles come up to the surface for picking.

Below: Pat and Jan's mobile burger van on the beach. This has to have been one of the most enterprising businesses I have come across. A sure example of when there is an opening…!

Above: The trailers head off
out to the sands, following a
well-trodden path. To venture
away from a known track can spell
danger on these treacherous sands.

Right: Transferring a bucket
of cockles into a bag. Hessian
bags have been replaced with
polypropylene ones, though the
size is about the same.

Left: A finished batch of cooked cockles. Much of the cockles these days go to Burry Port in South Wales for processing.

Below: A final reminder that the Flookburgh fishermen are also renowned for their cockles – as we shall see in a later chapter. (Photo courtesy of Jennifer Snell)

Oyster catchers are particular voracious feeders on cockles and one shot bird was found to have 140 cockles in its crop. Between 1956 and 1969 the Sea Fisheries Committee shot the birds, killing some 16,000 during the last four years. However, they found that their efforts were not having much effect upon the amount of spat, and shooting ceased.

So, as I sat upon my tuft of damp grass, the dog standing, nose in the air, gleaning the last whiffs of cooking from Pat and Jan's van, I was once again reminded of the fact that cockle fishing is one of the very few, possibly the only one, that has not changed much, even through the arrival of technology. Whereas any boat-based fishery has dramatically altered through the use of machinery and technical advances in both boat design and fishing gear, cockle fishing remains a manual task with craam, rake and jumbo. The horse might have been replaced by the diesel engine, but little else has changed.

Without waiting for the hordes of fishermen to return – I counted over 100 out on the sands – I was off, for it was time to observe the fishery in different pastures. However, it is worth mentioning that two weeks after I had visited the bay I read a small paragraph in the newspaper to the effect that, from midnight on 15 April 2005, the cockle beds in all of Morecambe Bay were to be closed because of dwindling stocks. According to Dr John Fish, the chairman of NW&NWSFC, 'a closure of all the cockle beds in the bay is necessary to protect the long-term viability of stocks.'

Paul was right in his aspiration that a closure was imminent. Although this fishery later opened, it was closed again in September 2008 and remains so. Furthermore, since the 2004 disaster, various legislation has been passed so that all pickers are now certified and gangmasters closely regulated. However, since that disaster, the words of Molly Malone's song are perhaps even more relevant:

Her ghost wheels a barrow
Through streets broad and narrow,
Singing 'Cockles and mussels, alive, alive, oh!'
(Traditional)

SIX

ARNSIDE AND THE MORECAMBE BAY NOBBIES

Silverdale, a few miles north of Hest Bank, and Arnside, on the southern bank of the river Kent, lie in the proximity of one of the most beautiful parts of the bay. Walking along the beach at Arnside, the promenade thronged with cars, the Albion Inn busy with afternoon drinkers, I capture some essence of what this place must have felt like a hundred years ago.

Visitors come to enjoy the solitude, the scenery and the social upmarketness of the place. Crossfield was a common name amongst the locals, the grocery shop of James Crossfield standing empty today. But the best-known Crossfield family were Arnside boatbuilders, and I had just returned from a walk along the beach to their old premises at Beach Walk to the west of the village, where they produced one of the most prolific fishing boats to be found in British waters a century ago. As a testament to their skill, many of these boats survive and sail today.

Also one of the best known of the old sailing fishing boats that worked off the British shores for shrimps, these were the Morecambe Bay prawners – also known as Lancashire nobbies or shrimpers or sprawners, depending on what preference you have. Why 'prawners' no one seems quite sure, for the only prawns they ever caught were off the Welsh coast! Some say it was conferred upon them by outsiders of the fishing industry. The name stuck, though. These boats habitually worked from the numerous harbours and bays of the west coast, anywhere between the Solway Firth and Cardigan Bay.

The nobbies, as I shall call them, seem to have originated from the Southport area where there were reported as being thirteen trawl boats in 1800, although these boats were very different to what we now regard as a nobby. Fish, it appears, were plentiful at this time – unlike now – and supplies of 'turbot, salmon, soles, oysters, shrimps and sometimes John Dory were repeatedly taken. Even though it is often mentioned in the annals of history, John Dory was an unpopular fish in the nineteenth century, unlike today when it has reached a high status amongst consumers. A bit like monk fish, which was fed to the crabs until their tails were found to taste like prawns in the twentieth century.

A view near Lancaster Sands, a print by F. Wheatley of 1787 showing small gaff-rigged vessels.

As seen in a previous chapter, it was reported in 1804 that a fishing boat called *Otter* (or so it implies) had arrived with the first catch of the season, bringing in '250 soles, plaice, cod etc.' Fishing was for many an active and important part of their life. 'Otter fishing', or in other words trawling, has been practised around the British coasts since Elizabethan times, but only in the early part of the nineteenth century did experiments lead to a development of the towed net which eventually gave us today's computer-monitored deep-water trawls that are proving so drastically effective in emptying our seas of life.

For example, Mr O. Williams of Llanidan on the Menai Straits introduced what he called the 'Torbay system of trawling' to Caernarfon Bay, and later Conwy Bay, about 1810. He obviously recognised that the men of Brixham were largely responsible for the substantial spread of trawling. However, centuries before him, in medieval times, the 'wondyrchoun' was an early example of a beam trawl that was notorious at the time for capturing fish. In 1376 this was recorded as being a net fixed onto a 10ft-long beam, weighted by stones, and three fathoms long and was thought to have developed from an oyster dredge. Trawling also occurred off the Flanders coast in the fifteenth century and this was prohibited in 1499, again in 1509 and in 1524. Trawling for fish has, since the first experiments, attracted controversy.

At the same time, herring fishing was fruitful in Morecambe Bay, and continued to be until it deserted the area about 1840. Maryport, in the north of Cumbria, was a significant herring port with many Lancashire and Scottish boats joining in for the season as a sort of unofficial holiday away from the labours of trawling. The practice of fishing away from home, taking the family with you on their summer holidays, survived until the 1920s. Somehow now it seems ridiculous to imagine fishermen uprooting and taking their kids off to sea, but considering the state of their housing in early nineteenth-century England, it was probably a holiday for all in its truest meaning.

Yet it was the shrimp for which Morecambe Bay became renowned as we have already discovered. The brown shrimp (*Crangon crangon*) was, as a rule, caught in small-mesh trawls which were towed by the smack-rigged nobbies. Similar boats, introduced in 1845 from Morecambe where they were built, worked beam trawls in the Solway Firth while being based at Annan. With fifty-one boats working in 1893, the fleet worked westwards with the ebb tide and returned on the flood. Several motor boats continue to use beam trawls there today.

The beam trawl essentially consists of a wooden beam – the length of which specifies the size of the trawl – on which two sledge-like irons are affixed, one at either end. From rings on the front of these are the bridle ropes by which the trawl is towed. The top edge of the bag-net is fixed to the trawl beam while the bottom edge is only attached at either end.

Along the bottom edge is the bottom gear, the means by which this edge travels along the seabed. Depending on the nature of the seabed, varieties of ground lines are available. Initially fishermen used a rope, and then this became weighted. Over

Crossfield's boatbuilding shed at Beach Walk, Arnside, with Francis John Crossfield standing on the right.

The Crossfield-built nobby, *Maud Raby*, sitting on a small trolley on the beach by the shed. (Photo courtesy of Jennifer Snell)

smooth sand this works quite well, but once the seabed becomes rough it just jumps up and is basically useless. Chains were then used to weight the ground line, and on rougher bottoms wrapped wire and looped chain or rubber bobbins threaded onto chain. In more modern times, large rubber discs or rollers are threaded onto the ground line to assist its passage over rocks and stones and give the trawl more momentum over obstacles.

Southport developed its own shank-net which was thought to be less destructive to young fish than were the beam trawls – a sign that some were recognising the effect that trawling was having upon fish stocks. Although not universal, the advantages of trawling were being weighed up against its depletion and destruction all over the country.

When ring-netting was establishing itself on Loch Fyne as an alternative to the traditional drift-netting for herring, many objectors eventually forced the Government to ban the method for a decade or more. They believed that trawling would eventually annihilate the herring fishery of the loch and recognised the ills of dumping dead fish into the sea.

This shank head was built up from a solid pitch pine foot – 10ft 6in long – with ash saplings bent to form a mouth about 15in high. Iron bars were fixed to the wooden foot to strengthen and ballast it, and the net was then fixed behind, much as a beam trawl net was, and the whole shank-net then towed along by bridle ropes.

Lancashire nobbies also caught pink shrimps (*Pandalus borealis*) in 18ft beam trawls in deeper water – in parts of the bay, the Lune Deep, Heysham Lake and off Blackpool

Another view of the *Maud Raby* at Arnside with Francis John on deck watching his rigger work up the mast. (Photo courtesy of Jennifer Snell)

The large nobby, *Red Rose*, at the launch at Arnside on 4 September 1895. This boat was one of the largest nobbies built at Arnside. (Photo courtesy of Jennifer Snell)

The launching at the Sluice from Wright's Yard in Shellfield Road, Marshside, Southport. (Photo courtesy of Len Lloyd)

Nobbies under construction at Latham's Yard, Crossen Sluice, near Southport, in the 1890s. (Photo courtesy of Len Lloyd)

Old and new style nobbies at Latham's Yard, again in the 1890s. The one with the cutaway forefoot is the obvious newer vessel. (Photo courtesy of Len Lloyd)

A nobby at Morecambe sorting and preparing the shrimps prior to boiling them up. Note the bobbins fixed to the bottom of the net's mouth. (Photo courtesy of Lancaster Maritime Museum)

A nobby off Arnside. By the look of the two ladies aboard and the hat of the man helping to hoist up the sail, these folk are off on a trip around the bay. (Photo courtesy of Lancaster Maritime Museum)

One of the oldest photographs of a nobby, this one being the *Titbits*, soon after her launch in Arnside in 1893. She was built for Harold Mount. (Photo courtesy of Lancaster Maritime Museum)

View of the deck of a nobby with the catch being sorted, Jack Mount being the fisherman behind. (Photo courtesy of Lancaster Maritime Museum)

Nobbies at Southport Pier, 15 August 1897 – time off from fishing for 'quality sailing' trips around the bay over the Bank Holiday weekend. (Photo courtesy of Len Lloyd)

A nobby drawn up the beach at Fleetwood. The man aboard appears to be mending the sail.

Nobbies drawn up alongside the pier at Grange-over-Sands. Trips over from Morecambe and Arnside were a popular outing and a welcome extra income for the fishermen during the summer season.

A shrimping nobby under shortened sail towing the trawl. The chimney of the boiler is clearly visible.

The small nobby *Cachalot*, CH10, upon the slip, showing her underwater lines which show her
to be an older vessel. The net on the slip consists of bobbins on a chain and the beam can be seen
behind. This boat still sails and is based in Conwy.

A small nobby in the river Lune under full sail and probably towing a trawl judging by the speed of the vessel and the disturbance in the water behind. (Photo courtesy of Lancaster Maritime Museum)

– rather than the brown shrimps that stayed on the shallow sandy banks. Locally these shrimps were called sprawns, and so the larger boats, as against the smaller shrimpers, soon became known as the sprawners. Superior sailing and fishing skills were needed to work these bigger boats (upwards of 34ft), so that a certain snobbery developed, whereby the shrimp fishermen, with their boats under 31ft, were regarded as inferior. To compete they enlarged their craft while the sprawners actually shrunk in size. The result was that by 1925 they all emerged as about the same size, between 32ft and 38ft, towing trawls of 9–11ft wide.

The first nobby as we now know it (forgetting those earlier boats from Southport mentioned above) seems to have appeared from the village of Arnside on the river Kent. Francis John Crossfield and his family began building there in 1848 (1832 has been suggested as the date of his first nobby though with no evidence), a tradition that the family retained until the late 1930s. It is generally believed that these craft developed from the earlier smacks into what are known now as the Morecambe Bay nobbies. A print of 1787, entitled *A view near Lancaster Sands*, shows a single-masted, clinker-built bluff-bowed vessel with a transom stern. Presumably boats like this worked the inshore waters of Morecambe Bay, and further away cutter-rigged boats were working off the Isle of Man.

From the mid-1860s onwards, the design of the nobbies developed. From the early transom stern, lute-sterned craft evolved in some areas. Another transition from this led to the square counter and eventually, and also finally, to the rounded counter. As has been the case in working boat development around Britain (and beyond) changes came about through the experience of fishers and boatbuilders alike, so that few boatbuilders built similar vessels, each putting their own stamp on their boats, each sharing their ideas but interpreting them in different ways. The first boats were clinker built until carvel building superseded the older method.

In the late 1890s the design altered somewhat to improve speed and handling, influenced by the racing yachts that some fishermen crewed aboard. Fishermen everywhere were keen to sail more powerfully, to tow bigger trawls, and speed was essential to get back to port as fast as possible to get the best price for the catch. At about the same time William Stoba came up with his ideas to improve the shape of the nobbies.

Stoba, born in 1855, was apprenticed with J. Gibson & Co., boatbuilders of Fleetwood, at the age of fourteen as a shipwright. Over the thirty-six years he worked for the company he progressed to designing boats for them. In 1905 he left that company and joined J. Armour, another boatbuilding company in Fleetwood, and began to put into practice his ideas. He suggested cutting away more of the forefoot and rockering the keel to produce a more efficient craft. This made them easier to steer and, when tacking, they were more responsive on the helm. The only disadvantage was that they would lay over on their side when drying out, for harbours were few along this coast, and remain so even today. Due to the way they lay on the sand, with

The lifeboat *Sir William Priestley* was built by Crossfields in 1934. Today this boat has been preserved at the Lancaster Maritime Museum, along with the whammel boat *Hannah*.

The nobby *Pastime* in the Menai Strait, *c.*1960. This was one of the first boats the author ever sailed in, her owner being a friend of the family. She eventually sank off South Stack, Anglesey.

The nobby *Hearts of Oak* sailing off the Isle of Man in 1994.

Venture of Chester, CH45, built in Arnside in 1920, sailing in the river Mersey in about 1998. Note the beam trawl over the stern.

the incoming tide some were swamped before they could righten. Despite this, boats of the new type were built in great numbers whilst the older straight-keeled boats gradually faded from popularity.

Crossfields became the recognised builder of renowned nobbies from his small shed on Beach Walk in Arnside. The company expanded in 1905 when Francis John's son – another John – opened a yard in Conwy. Ten years later two sons of George Crossfield – John's brother – opened a yard at Hoylake, though this enterprise only survived for five years. They continued building nobbies as well as yachts and even a couple of lifeboats for Morecambe in 1907 and 1934. When they ceased working at Arnside in 1938 they had built over 400 boats. Both Gibsons and Armour continued building nobbies, as well as the firm of Singletons, up to this time. Further south, close to Southport, some were built at the two yards of Latham and Wright.

Nobbies have a fo'c's'le ranging from about 11ft to 14ft in length, with a couple of bunks and a coal stove and not much more than 4ft of headroom at best. The cost of a new boat was, in 1912, about £60. Shrimps were normally boiled aboard during their dash home after ten hours or so of fishing, the catch being sold either upon the beach or from stalls. At the same time large amounts were sent by rail to the expanding city markets whilst some was potted, a delicacy for which Morecambe is renowned and which helped to develop the growing tourist market.

Another view of *Venture* as she motors into the Albert Dock, Liverpool, during the River Mersey Festival in 1998. Most of the nobbies sailing today have had engines installed.

The nobby *Naiad*, built by Gibsons of Fleetwood in 1894, languishing in Skippool Creek, on the river Wyre in 2009.

In time, use of the nobbies spread along the Welsh coast and into Cardigan Bay. Aberystwyth had nobbies working there, although many were owned by merchants from Liverpool. Motorisation improved their performance in the early 1920s and the first new build to be launched with an engine fitted appeared in 1925. Over the next forty years they continued working, as the hull seemed to work really well under power whilst the sails could still be used to reduce overheads.

However, their numbers did dwindle so that by the 1970s few worked. Many came out of the fishing industry and were converted to pleasure, and a healthy fleet often assembles for the Mersey and Conwy River Races each year. Today as many as fifty nobbies are said to be still sailing, many as members of the Nobby Owners Association.

SEVEN

FLOOKBURGH AND ITS SHRIMPS, HORSES AND TRACTORS

Humphrey Head, across the estuary of the river Kent from Silverdale, has the distinction of being the only sea cliff in North Lonsdale, the county region that borders much of the bay. In 1281, after King Edward I had commissioned Peter Corbet to slaughter all the wolves in Britain, this headland became the habitat of the last remaining wolf in the country.

A few years ago I had come to look for signs of cocklers, for I had been told this was one access for them onto the huge expanse of the Flookburgh Sands. In the grey light of an approaching winter dusk there was an eerie atmosphere about the place. The entrance onto the beach was blocked by a huge boulder and several stout-looking chain locks secured the gate firmly shut. Somebody obviously did not want people passing through here with tractors and trailers.

Walking along the cliff bottom, searching for the hidden holy well that the map suggested was somewhere here, and adding to the ghostliness of the place, I found a memorial plaque to William Pedder, ten years old, who met his end on these cliffs on 22 August 1857. Gazing out upon the sands, aware of the hidden quicksand and dangerous tides, I wondered just how many other unfortunates had lost their lives here in this inhospitable corner of the country. Wordsworth might have called these sands 'the majestic plain where the sea has retired' but they seemed anything but majestic. I am sure that it must have been the last wolf I heard howling amongst the darkened trees!

I drove a little west, into the tiny village of Flookburgh on the west side of the Cartmel peninsula on the north side of Morecambe Bay, and took the road south down to West Plain farm near the large holiday park. Access onto the beach was not blocked and I was able to walk the half mile along a fairly solid track, through foul smelling filth, across the marsh to the very edge of the sands. Empty cockle shells lay all about. The bay stretched out and beyond and I could just make out the bulk of

Flookburgh market cross from an old postcard. (Photo courtesy of Jennifer Snell)

the Heysham Nuclear Power Station. On a clear day Blackpool Tower is said to be visible. Humphrey Head jutted out a mile eastwards.

The beach was empty, the last of that day's pickers home now, feet in front of the fire. I had been told there were some twenty locals cockling, with a higher number of Polish migrant workers on the beds. The only signs of a flourishing trade were tyre marks crossing the stream where two bags of cockles lay in the shallow water. A blue glove lay in the sand, discarded maybe by a picker. The watery orange sun sank slowly into the horizon.

Further out I spotted the remains of a rusting chassis lying half buried. I crossed over a gully to look, suddenly noting a different squelching of the sand as I put one foot forward. In an instance that foot sank into the sand before I quickly withdrew it. Fearing quicksand, I made a hasty retreat. The reality of the danger of proceeding without any local knowledge was obvious. Walking back to the van I noted a high fence between the holiday camp and the beach, preventing the caravanners from getting on to the beach. I now knew why. This was no place to be in the falling dusk, or indeed at any time.

A few years later I was back, this time after the shrimpers. Shrimping in shallow water has persisted in various areas of Northern Europe for generations and again Flookburgh was the obvious place to look, lying right alongside the huge expanse of sand that is exposed at low tide. Rather than fishing with a boat, it was the horse and cart that was their everyday tool and this was used for all manner of fishing, not just shrimping – they were fishermen that never used boats! Cockles and shrimps accounted for their major catches, but they also set stake-nets for whitebait and flatfish, especially flounders which they regarded as being as tasty as plaice.

How long they had been shrimping with horses is unknown. Some say there were as many horses as folk that lived in the village at one time! It has been said that because Flookburgh was largely inaccessible before the twentieth century, no real market grew for their shrimps, athough they must have been catching them for the local market. However, it is known that hand-nets – sometimes called power or push-nets which were up to 6ft in width – were pushed through the water ahead of the fishermen. The nets were fixed to an ash hoop and the handle of pitch pine was 8ft long. Shrimp fishing in this manner was referred to as 'putting'.

The trouble with this method was that they could only go into shallow water for fear of being swept away. The shrimps generally were to be found in the fast flowing water of the channels, and in the dykes and gutters that flow into them. As the tide receded, the channels got smaller and so the concentration of shrimps in them higher. Thus horses were employed to draw larger nets through water that was a bit deeper. The net, of half-inch mesh and fixed to the back of a high-wheeled cart, was conical in shape and referred to as a shank, as already described.

The horse-drawn shanking carts dragged these nets through the shallow water, bouncing the shrimps off the seabed and into the nets as they progressed along the

A typical street view of Flookburgh at the turn of the century with cockle and shrimp carts parked up. (Photo courtesy of Jennifer Snell)

Harold Manning of Main Street, Flookburgh, the last fisherman to use a horse and cart for shrimping, seen here fully equipped for the sands. (Photo courtesy of Jennifer Snell)

Shrimping in progress off Sandgate, Flookburgh, with Jack Butler (on the left) and Jack Stevenson. (Photo courtesy of Jennifer Snell)

A Flookburgh fisherman attending to his stake nets out on the sands.

sandy bottom. These nets, some 12in shorter, were pulled parallel to the shore. The process was surprisingly technical and reliant upon the tide, as everything else on the Sands was. It was usually done downstream on the ebb which made things easier for the horse.

The weight of the catch was judged by the heaviness of the trawl rope. After stopping the horse, lifting and emptying the net into the cart, the whole outfit had to be turned around before the beam trawl was released just in case the net floated underneath the cart thus becoming fouled on the axle and wheels. If this was attempted with the net still out, it was likely that the whole cart would be overturned by the strength of the tide.

The horses came in all sorts of shapes and sizes, from the lighter-legged, half-bred, hunter to the ponderous Shire and Clydesdale sort. The lighter animals were favoured on the outgoing trip as they reached the shrimping ground faster, thereby claiming the best fishing position. On the return journey the heavily laden cart told on the tired hunter and, in the words of Jennifer Snell, 'the "tortoise" would be enjoying his warm stable whilst the "hare" was still plodding home'. They were all placid creatures, reliable and strong for the work, especially when they had to haul a full cart out of the wet sand. They spent many hours in the water and seldom had time to dry. Sometimes they were bought cheap at one of the local horse fairs as lame horses with an inflamed leg or foot, but after working in the saltwater for a few weeks this soon cleared up and the horse was sometimes sold at a profit a year later.

Shrimpers working at their maximum depth towing their trawls. It is easy to see what a dangerous occupation this was.

Two fishermen with their tractor-pulled cart, 1960s.

A typical beam trawl made with a rod of hazel.

A typical tractor with boxes of shrimps on the rear.

Shrimp pickers at work in 1926. Once the catch was boiled the tedious task of removing the shell was often regarded as women's work. (Photo courtesy of Keith Wallacy)

They learnt fast, these horses. The newcomer was at first easily panicked in wet sand though they soon became adept at recognising it. They had to understand the voice instructions from the fishermen whilst trawling, especially if an obstruction was snagged. They also learnt to follow their outgoing hoof prints on the way home and could sometimes 'smell' their way back to the safe shore. Occasionally they alone were responsible for fishermen finding their way home so that the men's lives depended on the horses.

Accordingly, they were well looked after and fed on a good diet of hay, oats and bran. They were also good at swimming with the cart attached when they were out of their depth, and many a horse safely carried his owner back across the channels in the sand. The bond between man and horse was very strong, and it was always a sad day when their horse finally had to be retired or, more tragically, died.

Death out on the sand was not uncommon for the fishermen. In 1857 ten fishermen drowned whilst returning home for Whitsuntide when their horse and cart overturned in a pool. In 1881 Margaret Sefton of Flookburgh, aged seventy-four, was overtaken by the tide whilst she was out picking cockles. Again, in 1912, three fishermen from Flookburgh were drowned. There were many more tragic deaths, and as we saw in chapter five working out on these sands was a treacherous occupation.

March to May and August to November was the best time for shrimping. Mostly it was trawled during the daytime, though sometimes the men went fishing at night, especially in the summer. The catch was placed in fish boxes aboard the cart and they headed home before the flood tide engulfed them. The shrimps were then skilfully

Carts on the marsh at Flookburgh in 2009. Most are homemade by the fishermen themselves, a practice seen throughout generations of fishing.

boiled as quickly as possible after landing, usually in the outhouse, and then carried inside to be picked at the big kitchen table. Much went to the Ulverston and Barrow markets and were sold in two or four ounce bags. Eventually hygiene regulations controlled the way they were dried; not on hessian sacks on the ground but in trays raised off the ground.

I was back again in 2009, buying Morecambe Bay potted shrimps in the newly refurbished premises of the Furness Fish & Game Suppliers Company. These premises were once the home of the Flookburgh Fishermen's Co-operative which closed in the 1990s. Les Salisbury, the owner of today's business, told me how he had been fishing since he was a young kid and that he had previously supplied most of the catch to the co-operative so it seemed sense to buy the factory and expand.

'Lots of folk go shrimping,' he told me, 'especially in times of recession.' An obvious referral to the practice that some go shrimping if they are out of work or on reduced wages. Some would be out later that day, he said, though with the farmers clearing ditches there was a lot of weed about and the shrimps were small. Small and big ones – locally called cowpikes – are not tasty enough to be used. 'I'll start about the end of August,' he added.

It seems that it was the Youngs Brothers who pioneered potted shrimps on a commercial scale in the 1950s, a time when there were still forty horses working in Flookburgh. They had opened a factory in nearby Cark, where picked shrimps were cooked in hot butter and spices and allowed to cool in wax-cartons. Potted shrimps

Morecambe Bay Potted Shrimps are today the biggest producer of the delicacy and supply markets throughout the country.

Smaller operators work from their backyards, such as this one in Flookburgh.

'Cockles Corner Shop', a reminder that Flookburgh isn't just about shrimps!

from the north side of the bay appeared in the markets. Before that it was down to individual fishermen to produce their own.

At about the same time, tractors appeared on the sands and even an ex-army DUKH had been used a few years before. Horses started disappearing so that within a few years they had all but gone. With tractors capable of pulling two trawls of a bigger 15ft size the catches increased, which, for a while, was grand. Then it became increasingly difficult to catch a big trawl and eventually the number of fishermen working the sands gradually decreased.

Today, as Les says, there are many who go out on a sort of part-time basis, and much of what is caught locally is processed and available under the Morecambe Bay Shrimp label produced in his factory. Here it is boiled, 'picked' by machine and potted with butter and spices before being sold through the Internet and the supermarket Waitrose, as well as to local hotels and a weekly market in London. Others, of course, still produce their own potted shrimps, such as I. & K. McClure of Flookburgh and Edmondsons of Morecambe. Some shrimp is picked and boiled at home. However it is produced, it remains a treat, and one only available from shrimps caught around the shores of the bay.

EIGHT

ULVERSTON AND THE *HEARTS OF OAK*

When one talks of Morecambe Bay nobbies, of which *Hearts of Oak* is one, one generally thinks of the renowned boatbuilding family – the Crossfields – whom we discovered in chapter six. Mention the town of Ulverston and you would be forgiven for not associating it with the nobbies. How wrong, though, for several were built there, including four or five by boatbuilder John Randall McLester.

Born in 1857 and brought up in Greenodd, some 2 miles from Ulverston, he was the last apprentice taken on by William White, a well-known shipbuilder working on the short Ulverston Canal that connects the town to the sea. The shipbuilders of the town were renowned for their exceptionally strong two-masted schooners that carried iron ore, an unforgiving cargo, all around the coasts of Britain and Ireland. However, by the time he had completed his apprenticeship there had been a sharp decline in the fortunes of local shipbuilding, and he moved to Barrow-in-Furness before returning to Canal Foot, Ulverston, to become landlord of the Bay Horse pub there, continuing to repair fishing boats on the beach.

Ulverston had had its short canal since 1796 and the Ainslie Pier since 1815, both of which were built to expand this iron ore trade. Boats of up to 350-tons burthen could enter. And busy it was. In 1841 some 944 vessels entered the lock. However, with the building of the Plumpton Viaduct for the railway line in 1856, the channels changed and silting caused serious problems to the shipping. Trade did continue after the building of what is known as the Collins Weir, the intention of which was to deepen the river Leven close by Canal Foot. The canal lock gates were finally concreted up in 1949. Today the site is overshadowed by the huge GlaxoSmithKline works, though without their financial help (and others), the restoration of the *Hearts of Oak* might never have occurred.

One local fisherman living there in the first decade of the twentieth century was Peter Butler, originally from nearby Flookburgh. In 1907 he was appointed the Over Sands Guide for the Leven Estuary, an ancient position going back several centuries

A Valentines postcard view of the lock gates at Canal Foot, Ulverston, *c.*1911. (Photo courtesy of Jennifer Snell)

Another postcard view of the entrance to the canal from about the point that *Hearts of Oak* first floated. The yacht is the *Nebula*, a frequent visitor, one that wintered in the canal and was the last vessel to leave in 1946. Note the netted–off swimming area across the beach. (Photo courtesy of Jennifer Snell)

The *Nebula* anchored in the lower canal basin with the Bay Horse Inn on the left of the buildings. (Photo courtesy of Jennifer Snell)

The SS *Oak* aground off Foulney Island in March 1907. The *City of Liverpool* has been helping to remove the ballast for easier floatation on the next tide. (Photo courtesy of Jennifer Snell)

whereby he must, when needed, guide travellers over the treacherous Morecambe Bay sands at low tide.

Fishing in those days in the estuary consisted mostly of shrimp trawling from the back of a horse-drawn cart in 4ft of water. In 1911, though, Peter Butler approached his friend and neighbour (and perhaps drinking partner) John Randall McLester to build him a boat for both fishing and 'excursion duties', by which he meant hiring himself and the boat out by the day. This would prove a lucrative sideline to his other work, for remuneration as a Guide was low apart from the large house next door to the pub and two acres of land that was included. By this time he had two daughters, Polly and Betty.

Thus the *Hearts of Oak* was born, a traditional Morecambe Bay prawner – or nobby – built stoutly from local oak felled and seasoned, so legend has it, by John Randall McLester himself and from which her name was derived. She was built on the beach alongside the Ainslie Pier at Canal Foot and she has been described as advanced for her time, especially in her modern spoon bow and, most unusually, by her centre board, fitted for fast sailing. Moreover she is regarded as having a particularly elegant counter stern. She measures 33ft overall, a beam of 10ft 6in and a 4ft draft. As it was, it transpired that she was the last vessel ever built and launched in Ulverston.

That launching took place on 8 June 1912 and seems to have attracted quite a lot of attention. His daughter Polly christened her that summer's day, smashing her mother's best glass decanter on the bow which was supposed to have been poured

The *City of Liverpool* was owned by the North Lonsdale Iron & Steel Company and is seen here laid up alongside the canal awaiting a buyer, *c.*1916. (Photo courtesy of Jennifer Snell)

Local fishermen and employees of the Ironworks sitting on the lock gates at Canal Foot, *c*.1920. Back row (left to right): Gregg Wilson (the lock keeper), -?-, John Wilson (Gregg's son), -?-, -?-, -?-, -?-, Bill Bird (fisherman), Thomas Gardiner (fisherman). Middle row (left to right): -?-, John Randall McLester (builder of *Hearts of Oak*), Peter Butler (Queens Guide Over the Leven Sands and first owner of *Hearts of Oak*). The rest are unknown. (Photo courtesy of Jennifer Snell)

John McLester and his family outside Sandside Terrace. (Photo courtesy of Jennifer Snell)

Launch day, 8 June 1912: bringing the boat down onto the beach on wooden runners. This was achieved solely by using manpower. (Photo courtesy of Jennifer Snell)

As the tide rose, the boat floated for the first time. Polly Butler is waving from the bow of the boat. (Photo courtesy of Jennifer Snell)

Peter Butler dressed as the Guide Over Sands standing in the entrance to a boat that appears not to be the *Hearts of Oak*. (Photo courtesy of Jennifer Snell)

Fully rigged and alongside Ainslie Pier at low tide, the photographer obviously carefully posed this picture with family and friends of Peter Butler aboard. Note the leg to stop the boat falling over. (Photo courtesy of Jennifer Snell)

Ulverston Swimming Gala at Canal Foot about 1914. *Hearts of Oak* is being used as a floating diving pontoon across the entrance to the canal gates. I am not sure that today's 'Health and Safety' people would be impressed with the method of getting aboard and the lack of railings on both the boat and the harbour! (Photo courtesy of Jennifer Snell)

On the sands by Ainslie Pier in the 1920s with Arthur and Lily Baxter standing. Note the trucks on the pier. The boat has had a wheelhouse added on by this time. (Photo courtesy of Jennifer Snell)

The *Hearts of Oak* as pilot boat assisting the SS *Ashfield* after she had gone aground on the previous day in September 1032. She had been unsuccessful in her attempt to reach Ainslie Pier. Unloaded by labourers from the Ironworks into horse-drawn carts using shovels, she eventually floated and sailed off. (Photo courtesy of Jennifer Snell)

Wartime at Haverigg when she was the pilot boat for the Millom Ironworks and named *Barbara*. (Photo courtesy of Jennifer Snell)

In the mid-1950s she was sold to Dennis Brown as a fishing boat and registered BW40. Here she is seen off Askam Pier in July 1964. (Photo courtesy of Jennifer Snell)

Denis Brown (owner) and Jack Allonby aboard a smaller vessel. (Photo courtesy of Jennifer Snell)

At Askam in 1960. Colin Dickinson is picking up the mooring buoy with the boat-hook whilst schoolboy David Standing looks on. (Photo courtesy of Jennifer Snell)

over. In the heat of the moment her father told her to 'give it a good bash', which she did, much to her mother's subsequent chagrin!

Thus the boat was worked locally, both at the fishing and the tripping. He often took the local children out during the summer holidays for a few hours although his best earners were fishing trips when groups of local businessmen hired him for days or even a week, often wanting to sail as far away as Kirkcudbright on the Solway Firth or the Isle of Man.

It was, in fact, in the Isle of Man where I first came across the boat back in the mid-1990s. By then the boat had had a varied life. Peter Butler died suddenly of cancer in January 1922 and the boat was beached outside his house. John Randall McLester was then acting as pilot for the few vessels inbound into Ainslie Pier to collect iron from the ironworks, and he suggested to the company that they purchase the *Hearts of Oak* to replace the existing smaller pilot boat. This they did and her name was changed to *Barbara* after Mr Tosh, the manager's, young daughter. The boat remained there until the closure of the ironworks in 1938, after which she was transferred to the Millom Ironworks to act as their pilot boat. During the Second World War she acted as an air-sea rescue craft and saved the lives of several ditched air crews off the Cumbrian coast.

After the war she was mainly used by two Haverigg men, Alec Mellon and Jack Taylor, until she fell into disrepair and languished, partly sunk, in a small tidal inlet close to the village. Still owned by the ironworks, Jack Taylor sought permission from them to renovate her, which they agreed to. In the mid-1950s she was sold to Dennis

Ted Gerrard holding the trawl rope whilst fishing for shrimps in the early 1970s. At the time she was registered as LR46. (Photo courtesy of Jennifer Snell)

Maryport, 1980, when she was registered as MT44. She was in a sad state when Jennifer Snell took this photo after seeing her for the first time. (Photo courtesy of Jennifer Snell)

Sailing in the Conwy River Festival in 1992. (Photo courtesy of Jennifer Snell)

Arrival at the GlaxoSmithKline compound in Ulverston at the end of 1999 after being transported over from Northern Ireland. (Photo courtesy of Jennifer Snell)

Outside the boatbuilding works of Waterfront Marine at Port Penrhyn, Bangor, in 2005 awaiting restoration.

Brown of Askam-in-Furness for use as a fishing boat and was registered as BW40 and returned to her old name. Dennis's brothers, Alf and Fred, nephew Ken, friend Jack Allonby and his son David all frequently sailed the boat, often over to the Isle of Man. When the creek that they moored her in silted up, they sold her in 1964.

The next owners were brothers Ted and Frank Gerrard of Morecambe, who re-registered her as LR46 (LR for Lancaster). They added a shrimp boiler, as was usual aboard the shrimpers, and fished her for ten years even though Ted regarded her as not ideal due to her high freeboard that made it difficult to haul aboard the nets. They fished as far as Lytham-St-Annes and the Dee and Conwy estuaries, often staying away for weeks.

In 1974 they sold her for £1,000 to John Dixon of Maryport. Again a change of registration – to MT44 – and she fished from Maryport for several years. It was here that Jennifer Snell, a local historian from Ulverston, first saw her after meeting Jack Taylor in 1980. She had first heard of the boat back in the early 1970s from John Wilson of Canal Foot whilst researching for a book *Ulverston in Times Past*. That first sighting was memorable because the boat was dismasted and sported, in her words, 'a broken-down cabin, rotting ropes and a discarded washing machine on deck'.

Eventually the boat was laid up ashore awaiting repairs. In 1982 she was purchased by Kenny Hall, a Wallasey classic boat enthusiast who sailed her off to the Mersey. She was de-registered for fishing purposes and he did some necessary work to her. However, she was eventually bought for £5,000 by Dennis Wright of Llandudno then later by Dave Pendleton of Conwy, who was in fact the owner when I first saw her, painted red in the Isle of Man.

Dave often sailed her in the annual Mersey Nobby Race, and I remember joining him aboard for the race in about 1996 when we came last. We had a good time aboard, even if the boat was badly in need of some serious work. That same year Jennifer Snell sailed aboard her in the river Conwy and told him of her twenty-five-year quest to find the boat and her desire to take her back to Ulverston. Two weeks later, though, Dave had to write to Jennifer to inform her that he had sold her to Stephen Clarkson.

Stephen sailed her over to Bangor, Northern Ireland, intending to sell his other boat *Vilya*. Seemingly he only just made it over for the boat leaked badly and the engine failed. When the sale of *Vilya* never materialised, *Hearts of Oak* was taken ashore and stored in a barn in Donaghadee. Stephen then found the website of the Ulverston Heritage Centre (of which Jennifer was chairman) and asked if anyone knew of the boat because he believed she originated from the town. Within a week and after several phone calls, he offered the boat to the Heritage Centre for a token £1. Money was quickly raised from the town council and the Ulverston Canal 200 Committee to pay for the transport costs, and *Hearts of Oak* arrived back in the town on 6 December 1999.

Thus began the hunt for money to restore her. The 'Hearts of Oak Boat Trust' was formed in 2001, made up of trustees and members, amongst whom are five members

(Left to right): Chris Thompson, Tony Walshaw, Alan Hind and Brian Scott, all of the Trust, talking to boatbuilder Scott Metcalfe aboard the boat in 2006. (Photo courtesy of Jennifer Snell)

Relaunch day, 12 June 2007, just over ninety-five years after she was built, at Port Penrhyn. Jennifer Snell and Jean Warton pour wine over her stem. (Photo courtesy of Mike Arridge)

of the McLester family, and this became a registered charity. Eventually their bid to the Heritage Lottery Fund was accepted and they were granted £49,500, which meant a further £10,000 or so had to be raised locally, both from local businesses and organisations and by the Trust staging various events.

In 2005 the boat was moved to Scott Metcalfe's Waterfront Marine in Port Penrhyn, Bangor, North Wales, and a full rebuild commenced. Over the next eighteen months much of the backbone was renewed, double sawn frames – moulded to 3in and sided to 2in – fitted along with new 1¼in nominal larch planking. Scott told me that his main shipwright, Joe Ormond, undertook most of the work, more than him in fact. Deck beams were iroko and 1½in Douglas Fir decking replaced in the original way, tongue and grooved. The original iron keel, with a slot for the centreboard, was found to be broken so a new lead keel of the same weight was fitted, without the slot and thus no centreboard.

Below decks the work was minimal, with a bulkhead to form a fo'c's'le with one bunk. Mast and spars in Douglas fir were made up by Scott, rigging and metalwork undertaken by John Duncan of Moreton on the Wirral, and sails in tanned clipper canvas from Steve Goachers of Cumbria. Thus, on 12 June 2007 she was craned back into the water almost ninety-five years to the day after Polly smashed the wine decanter upon her bows. This time Polly's eighty-four-year-old daughter, Mrs Jean Worton, poured the bottle of wine over the bow of the boat, with several descendants of the McLester family and members of the Trust watching.

In Jennifer Snell's own words, it was possible 'to see what a wonderful job Scott Metcalfe has made of restoring our old boat which to all intents and purposes was a write-off when she arrived at his yard in June 2005'. When I spoke to her after her first sail in the restored boat she seemed positively confident of the future. When I pushed her for a quote she hesitated before saying, 'It was almost unbelievable that the Trust managed to restore her as in the beginning it seemed an unattainable target. Sailing aboard her was a dream come true, especially for a land-lubber like me!' I typed as she spoke, and when I added the exclamation mark she laughed.

In August 2007 *Hearts of Oak* sailed to the Conwy River Festival where five of her keen crew were looking forward to trying her out in the Nobby Race which, unfortunately, was cancelled because of the weather. She then went back to Scott's yard to have an engine installed, after which she went back into the water and sailed back to Ulverston, the first visit back to her home town for over a quarter of a century.

'She sails really well,' said Scott at the time, 'although she probably needs a little bit more ballast and we're awaiting the topsail.' To date he has put some 1¼ tons of stone ballast off the beach, just another little sign of how he has tried to keep the restoration as traditional and close to the original vessel as possible. As Jennifer reminded me, she's a very different boat to the hulk I saw lying in the yard in Ulverston back in 1999.

However, like all these restored boats, she has to have a use once back in Ulverston to maintain her upkeep. 'She'll be a "living history" artefact for the people of Ulverston,' says Jennifer. 'We are planning seasonal themed short cruises for small

Two nobbies at Port Penrhyn – the *Hearts of Oak* lies against the Chester-registered nobby *Lassie*. (Photo courtesy of Mike Arridge)

Sailing off Bangor in August 2007 with Scott Metcalfe at the tiller. (Photo courtesy of Mike Arridge)

Sailing in the Duddon Estuary, Cumbria, in September 2008. (Photo courtesy of Brian Hampson)

groups with such subjects as history, natural history, geology and other interesting subjects. All led by an on-board local lecturer. She wintered in Ramsden Dock, Barrow, overlooked by three frigates intended for the Sultan of Brunei before going to her mooring at Roa Island for this summer (2009).'

Youngsters have not been forgotten either as the Trust intends to run workshops based on the boat which include showing how shrimps and fish were caught. I would have thought this would interest many an elder as well. She will be able to compete in both the Mersey and Conwy Nobby Races as well as participate in the various regattas around the Irish Sea. It will be interesting to discover if she really has paid her way by the time she's a hundred years old in a few years. But for now she is the result of one person's twenty-five-year obsession and determination to see her back home. And for this Jennifer should be praised and the fact widely recognised to encourage others to safeguard yet another little piece of our maritime heritage that is, otherwise, all too fast disappearing.

NINE

THE END OF THE ROAD –
THE FURNESS PENINSULA

Between Ulverston and Furness Peninsula the views over the bay are superb. At the tip of the peninsula are what were once a series of islands – Barrow, Walney, Roa, Foulney and Piel being the main ones – although today only the latter remains a true island.

Barrow Island has been absorbed into the Vickers shipyard, and is now attached to the mainland. Walney has its umbilical cord in the form of the 1908-built bridge from Barrow Island. Roa was once owned by the Furness Railway Company who built a causeway to what was the original port of Barrow until the development of Ramsden Dock in 1879. Foulney has its own causeway which was built to protect the harbour at Roa from silting up. Piel, still reachable by ferry from Roa, once had a pier, parts of which, on demolition in 1891, went to build the pier at Grange-over-Sands. Prior to these developments the inhabitants of the area used each island to land their fish and smuggle goods at various times.

Barrow could have become a great fishing port but instead it became an even greater builder of naval and merchant ships. Strangely, for a town lying not so many miles away from Fleetwood, Barrow shipwrights hardly built any fishing trawlers. Nevertheless, it seems worthy to add a few words about the town.

Like much of the bay, in the early nineteenth century there were few inhabitants and the census of 1841 shows that there were 152 folk in the vicinity. Farming, fishing and mining ironstone seem to have been the only preoccupations, though in 1836 it was described as 'the principal port of Furness for the exportation of iron ore and also visited for sea bathing' yet the port was little more than a creek. Some wooden boatbuilding occurred after William Ashburner set up a yard in 1847 and later moved to Hindpool. By 1850 the population had grown to 661, by which time the Furness Railway had arrived.

The same railway company then built the docks, the first of which opened in 1867, though the population in 1864 was already a staggering 8,000 and this was set to more than double by the end of the decade. Steelworks sprang up and the obvious

Above: Barrow had its own fleet of pilot boats, this one being the Number II boat *Albacore, c.*1910.
Built as a fishing smack in 1880, she was converted as a pilot boat soon after and eventually
broken up at Roa Island in 1924.

Opposite: Though Barrow was never recognised as a fishing port, boats nevertheless did use the
port. Here the Isle of Man-built Manx nobby *Gladys* is seen registered as BW13. Built in 1901,
she moved to Barrow in 1936 where the lug main was changed to a gaff. She remained at Barrow
for several years before being taken to Plymouth and, eventually, to West Ireland. She was brought
back to the UK in 2006 and is currently based in West Wales sporting her original lug rig.

Looking across to Vickerstown on Walney Island with two nobbies on the Barrow foreshore in 1902, one of which is registered at Barrow.

Another view across to Walney Island with nobbies and other assorted craft under repair on the Barrow foreshore in the 1920s.

Boatbuilder Joe Dowthwaite along the nobby *Nance* in his workshop.

The frame of the *Nance* as it is now in the Dock Museum at Barrow as an exhibit of how these boats were put together.

Barrow's fishing quay in 2009.

The converted ex-fishing boat *Sea Jade* on the beach at Walney with the Barrow bridge in the background.

Old boats end their days on the beach at North Scale, Walney.

The wreck of a wooden fishing boat with the 70ft trawler *Vita Nova* in the background. Affectionately known locally as the 'Roa Island Wreck', this boat is being brought back to life as a houseboat in a 'green project' by its owners Scott, Helen and Saffron.

The lighthouse at Rampside, otherwise known as the Needle, one of thirteen lights built for the approach to Barrow between 1850 and 1870. This was going to be demolished until locals had it listed as an historic structure.

result was shipbuilding. The *Punch* magazine of 1867 noted that the town had grown from 'the quiet coastal nest of some five score fishermen' into the workplace of 20,000 iron-workers, according to Bill Mitchell.

Vickers, the shipbuilding firm, started in 1897 and are still building naval ships today, including Britain's nuclear-powered submarines, within its huge shed that domineers the skyline. Other than that, there is a gas terminal from the Morecambe Field and a few fishing boats working from the quay alongside the Dock Museum, the latter being one place to learn of the town's history. In the channel between Walney Island and Barrow other fishing craft lie at moorings, many probably working part-time.

Inside the museum is the framework of the Morecambe Bay nobby *Nance* that was built in 1914 at Arnside. It was subsequently bought as a wreck by Joe Dowthwaite, a local shipwright who had served his apprenticeship with John Tyrrell & Sons of Arklow and had moved to Barrow to work for Vickers in the early 1970s. After his retirement he repaired wooden boats with his son Charlie in a workshop at North Scale on Walney Island. The museum itself is housed in one of the graving docks dating from 1872 which was built by the Furness Railway Company. It was in use until the 1950s. Outside the museum is Barrow's longest serving lifeboat, the Watson-class *Herbert Leigh* that was operational between 1951 and 1982.

BIBLIOGRAPHY

Ayton, R. and Daniell, W., *A Voyage Round Great Britain* (VIII vols, London, 1814–1825)

Cunliffe, Hugh, *The Story of Sunderland Point* (Sunderland Point, 1984)

Curtis, Bill, *Fleetwood – A Town is Born* (Lavenham, 1986)

Buckland, F. and Walpole, S., *Report on the Sea Fisheries of England and Wales* (HMSO, London, 1879)

Davis, F.M., *An Account of the Fishing Gear of England and Wales* (HMSO, London, 1930)

Holdsworth, E., *The Sea Fisheries of Great Britain & Ireland* (London, 1883)

Horsley, P. and Hirst, C., *Fleetwood's Fishing Industry* (Beverley, 1991)

Kennerley, Eija, *The Old Fishing Community of Poulton-le-Sands* (undated Lancaster Maritime Monograph)

March, Edgar, *Sailing Trawlers* (London, 1953)

March, Edgar, *Inshore Craft of Britain in the Days of Sail and Oar* (vol.II, Newton Abbot, 1970)

Mitchell, W.R., *Around Morecambe Bay* (Chichester, 2005)

Layfield, Jack, *Ulverston Canal & Construction of Collins Weir* (Barrow, 2007)

Lloyd, L.J., *The Lancashire Nobby* (Liverpool, 1998)

Pape, T., *The Sands of Morecambe Bay* (Morecambe, 1947)

Robinson, Cedric, *Sand Pilot of Morecambe Bay* (Newton Abbot, 1980)

Rothwell, Catherine, *Fleetwood – A Pictorial History* (Stroud, 2007)

Smylie, M., *Traditional Fishing Boats of Britain & Ireland* (Shrewsbury, 1999)

Wakefield, A.M., 'Cockling in Morecambe Bay' in *Pall Mall* (vol.XVI, London, 1898)

Other titles published by The History Press

Fishing Boats of Campbeltown Shipyard

SAM HENDERSON & PETER DRUMMOND

Readers will be fascinated by this insight into the history of Campbeltown Shipyard. Despite continuously diversifying its boatbuilding activities in order to survive the changing fortunes of the fishing industry, competition from foreign yards moved into a new dimension from the mid-1990s onwards. Today the empty buildings which once comprised the shipyard betray no trace of the hive of activity which once existed there, yet with former boats still turning impressive performances, the fishing industry will long remember the fishing boats of Campbeltown Shipyard.

978 0 7524 4765 0

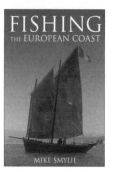

Fishing the European Coast

MIKE SMYLIE

No one knows when the first fishing boat set out to sea. Mosaics from the Mediterranean show vessels encircling shoals dating from the first century, although Egyptian tomb reliefs dated to 6000 BC show nets being set. Jesus, we are told, sailed aboard fishing boats on the Sea of Galilee around the early years of the first century AD, whilst Caesar noted that wooden boats were in use in Britain sometime after the Roman invasion. This book focuses on fishing boats of the last two centuries, and although the roots of some of the vessels may go back many generations, in the main they are still in existence in some form or other today.

978 07524 4628 8

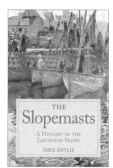

The Slopemasts
A History of the Lochfyne Skiffs

MIKE SMYLIE

The west coast of Scotland has its own peculiarities that have led to the development of certain unique boats. This fascinating book details the history of the Lochfyne skiff, which emerged from several generations of innovation and which resulted in one of the prettiest workboats to have graced the British shores. It was the last evolutionary stage in the era of sailing boats in the Clyde area, prior to the advent of motorisation in the first decades of the twentieth century.

978 07524 4774 2

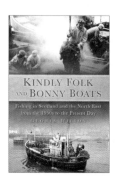

Kindly Folk and Bonny Boats
Fishing in Scotland and the North East from the 1950s to the Present Day

GLORIA WILSON

This book provides a pictorial appreciation of the boats and fishing communities of Scotland and North-East England from the 1950s to the present, making use of Gloria Wilson's unique collection of photographs. From attractive Scottish wooden-hulled craft to recent steel boats, and with many shore scenes including Mallaig herring port, Peterhead harbour reconstruction, fish auctions and fishermen net-mending and boat-building, this book offers a glimpse into a glorious bygone age.

978 0 7524 4907 4

Visit our website and discover thousands of other History Press books.

www.thehistorypress.co.uk